Future Humans

What will humans look like in one million years?

Thomas Mailund

Future Humans

Copyright 2020 Thomas Mailund

Table of Contents

Table of Contents		iii
1	Introduction	1
2	The Engines of Change	3
	Genetic Variation	3
	Enter Mutations	11
	Natural Selection	15
	Has evolution stopped?	21
	Changing Fitness Landscapes	26
3	Health	29
	Future Medicine	32
	Infectious Diseases	36
	The New Killers	53
4	Demographics	61
	Birth Rates	66
	Aging	81

	Sexual Selection	90
	Immortality	97
5	**In Our Natural Habitat**	**107**
	Urbanization	109
	Mobility	119
	Attracting Mates	122
	Pollution	126
	Ecological Impact	133
	Diminishing Resources	138
	Climate Change	149
6	**Going Forward**	**155**
	Colonizing the Oceans	156
	Colonizing Space	163
	Becoming New Species	169
	Unnatural Selection	175
	Birth of A.I.	180
	Ghosts in the machine	185
7	**Conclusions**	**189**
	Sources	191
	Acknowledgements	191

CHAPTER 1

Introduction

In 2012 and 2013, I assisted with a series of articles about human evolution from around six million years ago—when the last common ancestor of humans and chimpanzees lived—until the present. I have done research on the speciation of all great apes, including the split between humans and chimpanzees, so I felt well qualified to help with the first couple of articles in the series. The closer we got to the present of our species' evolution, I had to rely on papers by other research groups. Still, I had a firm grasp on the literature, and since the series was for a popular science magazine and a complete understanding of all the gory details wasn't necessary, I felt confident enough. After the series, however, I was asked to help with one more. For this last article, however, I was utterly unqualified. But not uniquely so. Everyone would be.

Just for the fun of it, I was asked what I thought humans would look like in a million years. No one knows, but it was a fun exercise to try to guess. One million years is an excellent period to play with if you are to speculate about our future evolution. It is long enough that you would expect us to evolve into a new species—the average lifespan of a mammalian species is about one million years. But a million years is also about four times as long as we have already existed as a species, so it should be clear that any predictions I made were completed guesswork. What I never expected would be the outcome of this guessing game was to be considered an expert in the future of human evolution.

Not a year goes by without me getting questions about this topic. It is not that I am interviewed frequently—journalists do not write about outlandish matters like this unless it is a very slow news day—but today I was asked yet again about our future evolution, and that triggered me to write this book.

In the following pages, I will give a slightly longer answer than I provide every time I am asked. And from now on, I will have a book to send people instead of writing long emails. I hope you will find it interesting.

CHAPTER 2

The Engines of Change

When I was asked to predict what humans would look like in a million years, the context was biological evolution through natural selection, not cultural or technological evolution. What I was actually asked was, 'how could humans evolve over the next million years?' and implicitly whether we are still evolving at all, now that modern medicine has blunted the razor of 'survival of the fittest'. In this chapter, I will argue that evolution cannot stop—change is always part of what a population will experience when new generations replace older ones. I will also argue that we are more, not less, equipped to adapt to our environment than we were in the past, whatever that environment may be.

Genetic Variation

The entire blueprint of your body, called your *genome*, contains all the genes that worked together to build you, from a fetus to an adult, and that are still working to keep your body running now. Your genome is a collection of DNA molecules that, if you laid them down in a straight line, would be two meters long. The building blocks of DNA, called nucleotides, are linked together into 23 pairs of long chains called chromosomes. The chromosomes come in pairs because you have two copies of what we consider 'the' human genome. You can think of this as the template for creating a human. You inherit one such genome from your mother and one from your father, each has 23 distinct chromosomes, so they combine to give you your 23 pairs. There are three billion nucleotides in each of the two copies. It would take you more than 95 years to count to three billion if you counted one number per second the entire time. If you need breaks, say for sleep, you could not do it in a lifetime. Forget about counting to six billion.

Each nucleotide is only a third of a nanometer in size. A nanometer is a billionth of a meter, so if you have three billion nucleotides, the two numbers combine to give you a meter long genome if you put your chromosomes end to end. You have two copies of a human genome, so your own genome is two meters long. That your genome, built from tiny, tiny molecules, is that long should give you some idea of how much information is encoded in it, and why it can code for the full complexity of your body.

The two genomes you get from your parents are not exact copies of their genomes. Each of the copies is created by copying different parts of their two genomes. So while you get one whole genome from each of your parents, those genomes are a mix of your grandparents' genomes. We do not pass exact copies of our genomes from generation to generation, but we mix them up each time, to create a new unique version for each child. This is why, ignoring twins, you never see identical siblings. If we did pass identical copies of one of our genomes to our children, then the chance that two children would get the same two copies would be one in four. The cut and paste of the grandparents' genomes increase the number of possible combinations to the point where it is virtually impossible to get two identical copies.

Your two genome copies look almost identical. Only about one in a thousand of the nucleotides in your genome differs between your two copies. We call the different versions *alleles*. Your two copies of a gene might be identical, so you have the same allele in both, or they might be different, so you have two different alleles. If you look at the entire human species, you will discover more differences than between your two genome copies. The more people you look at, the more variations you will find. There are seven billion people on the planet. At the time you are reading this, if you are reading it just a few months after I wrote it, it might be eight. The global population is growing *fast*. Each human has two genomes, so if there are seven billion people, the human species has 14 billion genomes. Whenever you look at a new genome, most of the alleles you see will be ones you have seen before, but a few would be new. For each new genome that you look at, you will find new variation until you have seen all that there exists in the species. You will find variation essentially everywhere in the genome. A few alleles are common in the species, but the vast majority are rare.

All genetic differences between us are the result of differences in alleles. The shuffling of alleles that happen when a parent passes on a genome to a child ensures that everyone's genome is a unique composition of alleles; it never existed before, and it will never exist again. All the alleles you have can also be found elsewhere; at least it is exceedingly unlikely that you have an allele that cannot be found in any other person, but you are the only one with this combination.

For most genes, it doesn't matter which alleles you have—they do not have any effect whatsoever. For others, one allele might make you a little taller while another will make you a little shorter; one allele might make your hair a little blonder; one might decrease your risk of cancer slightly. We call the combination of all the traits that we can observe *phenotypes*, and we call and the combination of all the alleles that affect them *genotypes*. Genes are not the only cause of phenotypes. For most of the differences we see between people, the genetic effect is smaller than the environmental effect. Smoking increases the risk of cancer much more than our genes—malnutrition in childhood matters far more for heights than differences in alleles. The environment we live in is likely to change dramatically in the future of our species, and this will have huge effects, but this chapter is about biological evolution and not environmental changes and their effect. I will speculate about our future environment in the following chapters.

You rarely see genes where one allele or another has a large effect. Typically, many genes will be involved in effecting a phenotype, and many alleles combine to make a noticeable effect. There is variation everywhere in the collective human genome, but we don't have all possible combinations of alleles. Our genetic differences sit on a background of mostly identical allele combinations. The shared background combinations are why we look so similar despite all the variation. Most alleles are rare, and individually they seldom change the phenotype by much; it is the common allelic combinations that give us the basic human body plan. You can think of the common alleles as a template for our species and the rare alleles as variations on that theme. If a different set of alleles were common, so the common combinations were also different, we would be another species. Chimpanzees have the same genes as we have, but for a few genes, they have one allele in nearly all of their genomes while we have an alternative allele in almost all of our genomes. Every genome will consist predominantly of common alleles, and the tiny fraction of rare alleles do not change the phenotypic effect of the common alleles by much. The common variants in chimpanzees define them as a species; the common variants in humans define us. The rare alleles only nudge us humans a little away from each other.

The frequency of the alleles in our species changes over time, and through this, the species evolve. Consider a couple with two children and consider a particular gene. Assume that one of the parents has different alleles for that gene. One of those alleles is lost if it is not copied to either child. Each of the alleles can end up in zero, one or two children. The probabilities are 25%, 50%, and 25%, respectively, because it is random which copy is passed on. So there is a 50% chance that both alleles survive and a 50% chance that one of them is lost. On a larger scale, consider the entire human population and the different copies we have of a specific allele. From one generation to the next, some people will have no children, and some will have many. When they do have children, some parents might pass on both of their alleles, while others will only pass on one of them. In genetics, we do not attempt to model the full complexity of this. Instead, we have mathematical models that let us work out the probabilities of how allele frequencies are likely to change as time passes. In the single-family example above, we saw a 50% chance that an allele would be lost in one generation. If we repeat the process through many generations, one allele will be lost with statistical certainty. For each generation that passes, the chance that both alleles survive is reduced by one half. There is 1/2 chance that both survive through the first generation, 1/4 that they make it through two generations, 1/8 for three, 1/16 for four, and so forth. Eventually, luck will run out for one of the copies. In the full population, we expect that all variation must eventually disappear as well, if not replenished.

In each generation, the frequency of an allele can take a step downwards toward zero, upwards toward one, or it can stay exactly where it is. The probability that the frequency stays the same tends to zero when we look at the process over many generations. If the frequency decreases in one generation, nothing in the inheritance process will actively work to raise it again. It doesn't get harder for the frequency to drop once it is closer to zero, it just gets more likely that it *reaches* zero—in which case that allele is lost from the population. The same goes for high frequencies. If an allele frequency is high, nothing prevents it from going higher until it hits frequency one. Then, it is the lone survivor, and all other alleles for that gene are gone.

The more people there are, the slower the change in frequency. For each generation, the change is the result of many people either passing a smaller, larger, or the same number of each allele on to the next generation. With few people, this process is highly random; with many people, it is less so. Consider tossing a coin. It can come up head or tail. We expect heads or tails to come up fifty-fifty, so if we toss a coin repeatedly, we would expect, on average, half of the tosses would be heads. With a single coin toss, half of them won't be heads; it is all or nothing. We cannot get the expected value. If we toss the coin more than once, we will get closer to see half of the results being heads. The more times we toss the coin, the closer the number of heads gets to be one half of the tosses. The more you repeat a random experiment, the closer the average outcome gets to the expected value. But at the same time, it gets less likely to be exactly one half heads. The more tosses in your experiment, the more outcomes you can get where around half of the coins come up heads, but out of those outcomes, the fraction where exactly half of the coins come up heads shrinks. Paradoxically, you get closer to the expectation but less likely to get the expected outcome.

The expected result of moving from one generation to the next is that the frequency of an allele stays where it is. It is equally likely that the frequency goes up as it is that it goes down, and the two probabilities cancel. The probability that it stays at its current value, though, is small; it is more likely that the frequency goes up or down. How far it moves depends on the population size. With a small population, as with tossing a few coins, the frequency is likely to move far from the expectation. With a larger population, just like tossing many coins, it is more likely that the frequency stays close to the current value (but less likely that it doesn't move at all). If the frequency changes, this is now the new expected frequency for the following generation. The more generations we look at, the less likely it is that the frequency stays the same because nothing in the process tries to return the frequency to its starting point. Unlike repeating the experiment with tossing coins, where each try uses the same coins and the expected outcomes are the same each time, the genetics experiment changes each generation as the frequency starts at different values. If an allele starts at 50%, we expect it to be near 50% in the next generation as well, but we also expect it to move somewhat up or down. If over generations, it reaches 25%, the expected value in the next generation is now 25% and not 50%. From here, it is equally likely to rise or fall in frequency. Now, it is equally likely to get to 0% as it is to go back to 50% again.

The probability that an allele will reach a frequency of one equals its current frequency. The allele, when it is at 50%, is equally likely to get to one as it was to get to zero. If it goes to 25%, it is equally likely to rise and fall in frequency, but there is now longer to go to one than there is to zero, and along the way, at each frequency it reaches, it is as likely to turn around and decrease as it is to keep increasing. From 25%, its chance of reaching 100% is only 25% while its chance of reaching 0% is 75%. If an allele fixates at frequency one, then whatever it codes for—big nose, hairy feet, etc.—is now shared by the entire species. The probability that an allele frequency never reaches either zero or one is vanishingly small, so if we wait long enough, all genetic variation we see today will be gone. All alleles will either have disappeared or have spread to the entire species. When all variation is gone, each new generation will be a copy of the one that came before. But this will never happen; while alleles are lost every generation, they are replenished by new mutations.

Enter Mutations

Mutations enter the species every generation. When alleles are fixed or lost, some diversity disappears, but new variation constantly enters the population to replace it. Based on recent measurements, we expect that about 60 new mutations per generation hit each genome. For a genome consisting of three billion nucleotides, this is a tiny number. The genome is large, and the mutation rate is low. But there are many human genomes. There are seven billion humans, so there are 14 billion genomes. With 60 mutations on 14 billion genomes, we expect 280 mutations at *every single nucleotide* in the genome if we consider the human species as a whole.

This back-of-the-envelope calculation doesn't tell you that every nucleotide in the genome is hit in each generation; it is only the average per nucleotide. Some are hit many more times, and some are hit fewer or zero times. There are more types of mutations than those that modify a single nucleotide. Those are less likely to appear multiple times; some will almost certainly be unique. However, your intuition should be that the human population size is so large that the mutation process adds variation everywhere in the genome all the time. Each individual genome is changed very little—it will still have all the common alleles that make us human—but every genome gets a few changes that evolution can play with. Because there are billions of human genomes, each generation tries out all possible variations on the basic human theme. You should not hold your breath waiting for X-Men mutations; all possible mutations have already been tried multiple times just in the last generation, and we have not observed any superpowers.

If mutations hit everywhere in every generation, we cannot fixate an allele. A mutation can reach a frequency where all mutations that occurred before it are lost, and all remaining variation is from mutations that happened after it. It can be the last survivor of all that came before it, but fixation in the sense that it has frequency one is impossible. Still, an allele can be at such a high frequency that it doesn't matter; only a tiny fraction of humanity will have an alternative allele. So we can pretend that we have replaced one variant of a gene with another in the entire human species.

Since the probability that an allele reaches fixation is the same as its current frequency, the probability that a single new mutation fixates is the same as the frequency of one single genome in the entire population. With 14 billion genomes, a hopeful new mutation has a one in 14 billion chance of fixation. So naturally, the vast majority of mutations are lost before reaching high frequencies. We can calculate how many mutations make it through to fixation from the one in 14 billion chance. If each of the species' 14 billion genomes is hit by 60 new mutations each generation, and each of those mutations has a one in 14 billion chance of fixation, then naturally, we expect 60 of the new mutations to fixate. It is not just a fluke that the number of alleles fixed matches the number of mutations. It is a general property derived from population genetics theory.

We can work out from genetics models that if a mutation gets fixed, the expected time it took to get there equals the number of genomes in the population, so we expect there to be 14 billion generations between when the mutation appeared, and when it reached fixation. The human generation time is about 30 years, so this is 420 billion years, which is a very long time indeed. The age of the universe is less than 14 billion years. Of course, the expected time is an average and not the time that all alleles would take to fixate. Some can reach fixation much faster, while others will take much longer. Still, with 7 billion people, we don't expect that any mutation would have made it through the entire species. And yet, some alleles must have reached substantially different frequencies in some populations than others. We can immediately see this from ethnic differences. Some allele frequencies have changed surprisingly fast compared to our expectations. One reason is that not all fixations are the result of random frequency changes from generation to generation. Adaptive changes, those that benefit us by adapting us to our environment, will move much faster to fixation (see the next section). Few mutations are adaptive, though, and must be the result of random fixations. As far as we can tell, the vast majority of alleles have no adaptive effect, but *are* moving randomly. Clearly, something in the calculations must be wrong.

Any calculation based on a human population size of seven billion will be wrong if we look more than a few decades back in time. We only recently grew to the population size we have today. We reached one billion at the beginning of the 19th century, three in the middle of the 20th century, and we have reached seven billion two decades into the 21st century. If we go back ten thousand years, to the beginning of agriculture, there were only one to five million humans on Earth. This is an explosive growth, and any calculation based on 14 billion genomes now does not work if we extrapolate to the past. Furthermore, the theoretical result that says that the time to fixation equals the number of genomes in the population assumes that everyone is equally likely to mate and equally likely to mate with anyone else within the entire human species. This clearly isn't true. Adjusting for this shortens the transition time from when a new mutant appears to the time it has spread to all humans (assuming it does this) by orders of magnitude.

The average transition time from mutation to fixation in the past is estimated to be in the tens to hundreds of thousands which corresponds to a population size of only around 10,000 humans. Such an adjusted population size is called an *effective* population size. It is a parameter we use in genetics models to compensate for demographics. We can estimate this number from genomic data, and then use it in a model that assumes a constant size population with no demographic structure, to draw conclusions about the real population. For past humans, the number is 10,000 individuals. It will be much larger for the future. We have a substantially larger population size and we mix more when it comes to mating. The transition time from low to high frequency will not be hundreds of billions of years, but it will be vastly slower than in the past. How much depends on how we mix but it will be measured in millions of years.

If we count mutations we do not have to adjust for demographic parameters as we have to when we consider the rate of frequency change. Each genome gets a number of mutations and it is the number of genomes in the species that count, not some adjusted size. The population size is expected to grow over the coming decades, but the growth is slowing down. We will be billions more before the growth levels off, but not orders of magnitude. So I am simply going to assume that we stay at the current population size. If we extrapolate from our current population size into to the future, the large population size means that we expect mutations in the combined species' genome to affect every position in our DNA, giving us an immense genetic variability. It is a situation very different from when the population size was merely in the millions and received orders of magnitudes fewer mutations—each genome got the same number of mutations but there are a thousand more genomes now. You could wait hundreds of generations for any particular mutation to appear. Today, we see that mutation multiple times each generation.

The number of alleles that fixates each generation is still the same as the per genome mutation rate—where with fixating we mean they reach the highest level they can reach, considering that new mutations occur all the time. The time it takes for a new mutation to reach fixation will be glacially slow compared to the past for alleles whose frequency drift randomly through the population, but if there is *selection* for an allele, then the story changes completely.

Natural Selection

Nowhere above did I invoke natural selection. We will see an evolution of our species' genes just from the random turnover of alleles—at a pace that makes geological time scales seem rapid, admittedly. Over time, this will change what our species look like, but much too slow to be relevant for the next million years. If this were the rate of change, there would be practically no difference between now and then. If we add selection, the time scale at which we see alterations shorten dramatically.

Whenever an allele is more likely to be copied to the next generation than alternative alleles, we say that the allele is selected for. There are many ways that this copying bias can manifest. The phrase 'survival of the fittest' suggests a 'red in tooth and claw' fight for survival, but this is not the right way to think about natural selection. The important aspect of selection is that one allele has an advantage over others. When we talk about the fitness of an individual, we mean how more likely it is that he or she to have an above-average number of offspring. It is fitness in this technical sense that drives selection: A fit individual leaves more descendants. The alleles that make him or her fit are therefore copied into more future genomes as we move forward through the generations. In a gene where one allele is selected for, the other alleles are selected against. The bias in one allele's favor acts against the alternative alleles; if one allele is more likely to be copied, those that compete against it are at a disadvantage. The two are dual to each other. We say that alleles that are selected for, are under *positive* selection, and those that are selected against, are under *negative* selection. Those that are neither selected for nor against are called *neutral*.

Survival is important, obviously, since not surviving prevents you from having more children. Still, survival is only one of many aspects of fitness. How likely are you to attract a mate? Without offspring, your genes are out of the gene pool. Do you want a small or a large family? With few offspring, you have fewer copies of your genes competing to rise in frequency against all other alleles in the population. Do you care for your offspring and make sure that they survive and prosper, or do you let them fight for themselves? Caring takes resources that you could use to have more children instead, but unless your children also pass on their genes, your genes have met a dead end. Anything that increases the chance that your genes are passed on to future generations makes you fit in the evolutionary sense.

When there is selection for an allele, that allele's frequency behaves vastly differently from neutral alleles. The frequency of a neutral allele moves as a random walk; each generation it is equally likely to increase or decrease in frequency. It is this aimlessness that makes it highly unlikely that a new mutation gets to a high frequency and makes the transition from low to high frequency so slow if it does get there.

As we saw earlier, we can think about genetic changes in terms of changes in allele frequency, but these changes are driven by random events where, in each generation, each allele tries to get copied. It is a random process for each allele. For any individual allele, the outcome is highly stochastic, but for the population at large, the change in frequency is more predictable. Compared to a neutral allele, where the frequency is equally likely to rise and fall in frequency, alleles that are under selection are biased in one direction or the other. For an allele under positive selection, the bias makes it more likely to rise in frequency. The more copying events we see each generation, the stronger that bias manifests. For every single copy, we still see a highly random process, but with enough allele copied each generation, the more likely it is that the change in frequency matches the bias in the selected allele's favor.

Since it is the total number of copies that matters for how

strongly the bias influences the next generation, we can make a few observations. As a general rule, the higher the selected allele is in frequency, the more copies we have, and the less random the process is, but this is not independent of the population size. In a small population, it takes fewer copies to reach a given frequency, and fewer copies mean more randomness. Hence, the frequency at which selection takes over from randomness depends on the population size. It is another case where our currently large population size makes us different from humans in the past. We have a larger population, so, at any given frequency, selection works stronger than it did in the past. Regardless of population size, at very low or very high frequency, the process is still mostly random. What constitutes 'low' or 'high' frequencies depends on the population size, but if we make them extreme enough, we reach a point where there are few copies of an allele or its counterpart. At extreme frequencies, the allele frequency in the next generation is determined by a few individuals' chances of reproducing, either those having the allele (when the frequency is low) or those without it (when the frequency is high). Either way, the changes in frequency will be small—if you add one or two extra copies in a generation and you start with 100 copies out of a billion, it doesn't change the frequency much. The randomness mostly swamps the statistical bias. A single copy of a positively selected allele is still highly likely to go extinct. It is the total number of alleles—not their frequency—combined with the strength of the selection, that determines how likely it is that the allele count gets a moderate frequency where the bias starts to matter.

Once the number of copies gets high enough, the process is almost deterministic. The allele will very quickly sweep through the population to fixation. This sweep happens at a speed that depends on the size of the selection bias but not the size of the population. The slow parts of this sweep through frequencies are at low and high frequencies, where the process is determined by a few copies. The slow parts do not depend on the population size, and the fast sweep doesn't either, so the time it takes an allele under selection to spread through the entire species is not slower in the future, even though the population is larger. It will take much longer for a neutral allele to fix, with the larger population size, but we will adapt just as quickly as we did before. Except, with the larger population size, we will always have the variation we need to adapt when our environment changes. The time we have to wait for an adaptation is the time we have to wait for a variant to enter the population and then the time it takes for it to sweep through the population. Our larger population means that we do not have to wait for new mutations, and the sweep time is just as fast as before, so if anything we expect our future evolution to be more effective than before.

The same process that drives positively selected alleles to fixation also eliminate alleles under negative selection—the two processes are, after all, the same, just viewed from different angles. Negatively selected alleles can rise to high frequencies and even fixation, but they have to be very lucky and very quick. You might think that if it takes neutral alleles such a long time to reach fixation, it will take negatively selected alleles even longer, but this is not the case. It is much harder for such an allele to make it through the entire population. It will have to be lucky almost every generation since selection is actively trying to push it down in frequency. This luck cannot last long, so should an allele under negative selection make it through the entire population, it has done so in a relatively few generations. With a larger population, more individuals might copy the deleterious allele, and in each copy, there is the selection bias. With more copies, the effect of the statistical bias is stronger, so with a larger population, we should see fewer alleles under negative selection reach high frequency and fixation. In a small population, like our species in the past, negatively selected alleles can float around in the population at moderate frequencies: the fewer people, the higher frequency before selection kicks in. With a larger population, we expect to keep these alleles at a lower frequency. In the future, we should see fewer of these alleles at a moderate or higher frequency. We will still have them at low frequency. With the species-wide mutation rate, this is unavoidable. With more genes copied each generation, there are also more chances that a deleterious allele makes it to the next generation. At a frequency in the population at large, though, selection will be better at preventing alleles under negative selection from spreading.

Selection, both positive and negative, will be more effective in the future than it was in the past. With a human population size in the billions, each allele is expected to enter the population each generation and at a high number of copies. This means that adaptive, i.e., positively selected alleles are unlikely to go extinct and will experience enough of the bias to get to a frequency where the process is almost deterministic. Then the allele will sweep through the population. We can expect evolution to adapt us to our environment faster than ever before. Deleterious alleles also enter the population in large numbers. Since it is easier to destroy than create, also for genes, most mutations with a phenotypic effect are destructive, so many more deleterious alleles enter the population than positively selected alleles. With the larger population size, though, they are kept at low frequency. The larger population size means that our future evolution is less random than our past evolution was. Our fitness will increase as we remove deleterious genes faster and adapt quickly.

Has evolution stopped?

If we ignore selection, alleles will still get fixed randomly, and new mutations will still enter the population. Unless we can stop mutations entirely—which seems incredibly unlikely—or humanity goes extinct, this process will always continue. The composition of alleles in the species will slowly change as some variants increase in frequency, and others decrease. Some mutations will be lost, but new ones will continually show up to replace them. The high rate of mutations ensures that we will always have the same alleles in the population, but this does not imply that the population never changes. It is the combination of alleles in many genes that determine what we are, and a rare variant here and there does not change the vast collections of common combinations in the species. With this random process of shifting allele frequencies alone, we will slowly and aimlessly drift until we would be considered a new species. Even with no selection, we would still evolve. Just very slowly. With selection, our evolution will progress considerably faster.

For selection to matter, we must have something to select *for*. If no new mutation provides us with a substantial selective advantage, then we only have the slow random genetic drift to change our genomes. But are we at perfect fitness, from where no improvements are possible? Hardly. For one thing, consider diseases that are partially genetically caused. We know many alleles that increase the risk of various diseases, some substantially and some slightly. Unless there are selective reasons to keep these alleles around—which there might be since genes can have more than one function—evolution should be able to get rid of them.

With modern technology, we no longer need to outrun predators—at least most people don't. We protect and feed those that cannot do so themselves. Modern medicine alleviates the impact of infectious diseases, something that has been a considerable driver of adaptation. We are not at the point where infections are harmless, the COVID-19 pandemic testifies to that, but infections are far less likely to kill us today than in the past. We have removed many of the drivers that guided selection before, and for those areas of the Earth where these modern advances are still not widely available, they soon will be. While new technologies are not evenly distributed, they spread through all populations, eventually, if the economic and technological trend that humanity is currently on continues.

If these selective forces are gone, we can then ask ourselves if there is still an opportunity for adaptation. These environmental factors might no longer require (or enable) selection. I have often been asked if evolution has stopped because our environment is no longer as hostile to us. In the world today, we have eliminated many of the dangers of the past, but the changes that removed the old selective pressures were changes to our environment. We now find ourselves in a world we didn't evolve in. We are not adapted to the modern world; there hasn't been enough time. The traits that evolution will select for in the future are different from the traits it selected for in the past, but there will still be selection.

Not everything is simple, or even possible, to evolve. Single mutations do not drastically change us. You will not see a mutant born with wings. That cannot happen. Many genes need to change to transform arms into wings, and for one genome to mutate at all the right places is impossible. It doesn't mean that we cannot evolve into a species with wings. Birds and bats managed to turn arms into wings, so there is, in principle, nothing that prevents us from doing the same. All it requires is that the right variants enter the species, and then they get shuffled around—recall that in each generation, we create new genomes by mixing our parents' genomes—until the right combination is found. This combination will never happen by pure chance, though. There are too many possible combinations that we can create genomes from, that picking just the right one is statistically impossible. To get to the right combination, we first need to get all the relevant alleles to high frequency. If they are sufficiently frequent, it is more likely that a new genome will have all of them. Getting them to high frequency requires luck, which won't be enough if there are many of them, or it requires selection. For selection to help us, we need selection for *each* of the alleles individually; we want all of them at high frequency. Maybe wings are incredibly advantageous, and maybe once we have wings, we will see selection for improving them, but we won't see that selective advantage before we have something that resembles wings. We need selection for each allele on its own, in the context of other common alleles. Natural selection needs to *see* the selective advantage there is in an allele before it will increase its frequency. Each of the alleles for creating wings must improve upon the phenotype caused by the other alleles in the population. Otherwise, they will not move from low to high frequency. To build wings, we must raise the frequency of the alleles one at a time, where each allele slightly improves on the existing body plan, moving us closer and closer to wings. There is no plan in this; selection won't help these alleles because it knows that we are working on building wings. Each allele has to bring something to the table on its own to be selected for. Each allele doesn't need to be selected for in the whole process. When some of the alleles are fixed, the core human genotype has changed and an allele that might have been neutral or selected against is now positively selected for in the context of the new common alleles. The process is gradual: some changes enable selection for other changes that enable selection for yet other changes. At each step along the way, though, we need selection.

We build fitter phenotypes by gradually improving on the existing phenotypes, a few alleles at a time. We do not make big jumps. To get from arms to wings, we need to go through many intermediate steps. If there is selection against one of those steps, we are unlikely to take it. Wings might be better than arms in the future, but to get wings, like those we know from birds and bats, we need to lose our hands. The steps along the way might give us less useful hands before we get the benefits of proto-wings, and selection would steer us away from that path. Selection gradually improves on what we already have; it does not make wild jumps from one form to another. If we have major changes in the future, they will evolve slowly and will follow paths where each step improves on the previous state. We might see radical changes in our future, but they will take an extremely long time to come about.

Evolution cannot take the direct route to an improved state; we have to take a detour, so each step along the way increases fitness. If all paths to a new phenotype require many deleterious steps, that phenotype is out of reach from us. There are many phenotypes found in nature superior to what we have. The octopus's eye is better than ours; its photoreceptors point outwards while ours point in the wrong direction, with nerves in the way of the light they are there to catch. It gives us our blind spot and generally makes them less efficient. Both are eyes, so you would think that we could get from one form to the other. But to do that, we have to restructure our eyes completely. To get the octopus's eye, we would need to lose our own and then build up the new kind of eye. The destruction of our eye will be selected against. There might be selection for creating the octopus's eye once we start from no eye, but getting to that point will reduce fitness each step along the way. That eye is out of reach from our evolution. The direction that our evolution took in the deep past led us to a point where we cannot go back and try an alternative route. Each step evolution takes us cuts us off from places where other paths might have taken us. In nature, we have features that we know can evolve, but they may be unreachable for us. Evolution cannot take us everywhere.

We don't have dramatic evolutionary changes in our future, at least not in the next million years. But we will still see changes. Evolution has not stopped; there is still room for improvement. There will be some selection forces we have always had, like resistance to diseases and attractiveness to the opposite sex, and some new directions because our environment has changed, and we haven't adapted to that yet.

Changing Fitness Landscapes

The metaphor 'fitness landscape' is used to describe directions of selective advantages and disadvantages. Imagine different genotypes (genomes and the alleles they contain) on the x-axis and their particular selective advantage as hills and valleys on the y-axis. The idea with the metaphor is that changes in genotypes can move us through the landscape, and selection will try to lead us uphill. Mutations that drive us uphill will be selected for, and the entire population will climb uphill; variations that have negative selection will take us downhill, and the population will avoid going in that direction. Selection will ensure that we approach the peak of the closest hill. The process is greedy, and it doesn't think ahead; it will always try to take uphill steps and never downhill steps. Thus, we are not guaranteed to find the tallest peak in the entire landscape—that might require moving down through a valley before we can climb another hill. If we are at a local optimum, we are staying there, even if there are better places somewhere else. Since the octopus's eye is better designed than ours, we are clearly not at the highest peak when it comes to eyes. Still, because we would have to destroy our own eye to change it into the octopus design, because we would have to descend deep into a valley before we can climb the higher peak, selection is never going there.

Generally, passing valleys is hard. A mutation that takes you down into a valley will be selected against. In a large population, the selection is strong, and we will tend to stay close to a peak; we will select strongly against moving into valleys. In a smaller population, it is easier to enter a valley so you might explore more of the landscape—if you do not go extinct, which is a considerable risk with a small population. The size of the population that lets you explore valleys also makes it less likely that you climb another hill, though. We are usually better off staying at the peak we are currently trying to climb.

If the fitness landscape changes, the hills and valleys move. Consequently, the hills our genomes were trying to climb might no longer be there. If modern living has removed selective pressures that previously rose as hills for us to climb, making these hills flat enough that our genomes will no longer climb them, or even making them valleys that our genomes will try to escape, we are in a new fitness landscape. But we are still in a fitness landscape. Selection only stops if the landscape is flat, or near enough that it is indistinguishable from flat. Otherwise, there will be other hills that we can climb, and I will suggest some in the following chapters.

If someday in the distant future, we reach a static environment and a peak in the resulting fitness landscape, our evolution will stop. Selection will still be present, but it will act to keep us on our fitness peak, actively preventing us from changing further as a species. Neutral alleles would always drift randomly up and down in frequency, but we would see no further adaptation. I don't expect this ever to happen, though. There will likely always be factors outside of our control. A new virus, for example, we do not have medicines or vaccines against, could put pressure on our immune system and force us to evolve. We might get stuck on the peak of a fitness hill, but it will not be for long. Eventually, the landscape shifts again, and we will start chasing the next peak.

CHAPTER 3

Health

Our environment has changed dramatically over the last ten thousand years and consequently so has the fitness landscape for our genes. The environment we live in today was not created by nature, but by ourselves; it did not gradually emerge while we evolved along with it, it appeared in an evolutionary blink of an eye, and we have not had time to adapt to it. From we emerged as a species and colonized the world, we have adapted our environment to fit our needs rather than adapted ourselves to fit our environment. We rapidly change the world we live in, but not perfectly. We did not shift the fitness landscape, so we are now on a tall peak; we merely changed it. Our primate ancestors climbed peaks in the old landscape, and some of those peaks remain for us to ascent further, but we have also created new ones, hills that never existed before; we are only at the foothills of those, and our climb is just beginning.

Many of the environmental factors that potentially created

a selection gradient in the past have now lost their effect. In changing our environment, we have flattened some hills and valleys in the fitness landscape. Antibiotics mean that we do not need as powerful a built-in immune system adapted to fight bacterial infections. Being physically weak has little disadvantage in a society where technology made the need for raw muscle power obsolete. Poor eyesight, such as myopia, is easily corrected with glasses, removing any fitness disadvantage there is to it. Our ancestors experienced selective forces here, but we no longer do. They, however, saw little selective advantage in dealing with diabetes and obesity. The overabundance of food that we have readily available in much of the world is posing a health risk, but this was never an issue in the past; starvation was, and adapting to deal with starvation might have left us particularly prone to lifestyle diseases. In many ways, we are ill-adapted to modern life; it came so rapidly that we haven't had time to change, but now the selective forces are there, and they will guide our future evolution.

Our fitness landscape is remarkably different from the fitness landscape early humans evolved in because of our technology. Our technological innovations have kept changing our surroundings at an exponential rate since the Stone Age. Back then, millennia could pass between technological innovations, but through history, we see a steady acceleration in the speed of inventions. When we look back 50 years, we see the same amount of change as when we look at the 100 years before then or the 200 years before that again. If we look forward 25 years, we are likely to see as many changes as we saw in the last 50 years. Accelerating technological progress does not imply accelerating changes to our fitness landscape. Not all changes will affect the landscape equally. Agriculture changed our diet dramatically, but I doubt that the invention of smartphones had as profound an effect. Still, current and future technology will affect our evolution, and that is what this chapter is about.

I will not try to predict specific future technology. It is, pure and simple, impossible. People from 1920 could not possibly predict the technology we have in 2020. With exponential growth in technology, it is likely to be twice as hard for us to predict the technology of 2120. I am certainly not going to make guesses for 1,002,020. I would hesitate to guess which technology we have in ten years. I would be confident in guessing culture trends for the next few decades—although I wouldn't bet money on it—but over centuries, I would undoubtedly get more wrong than right. I will not make predictions such as 'we will get larger thumbs because we are texting'. Honestly, I have been asked whether we would in a radio interview once, and it was not in jest. The question assumes that we would be texting over evolutionary time. We weren't texting 30 years ago, and I doubt that we will be 30 years hence. Any prediction based on a single technology will certainly be wrong. Still, if we consider general trends rather than specific changes, we can dare to speculate.

I will make some assumptions based on current trends and assume they continue far into the future and that a technological or cultural revolution does not dramatically change the direction of these trends. It is the best I can do. Since technology and culture change many orders of magnitude faster than the pace at which evolution modifies us, I have to be conservative in my assumptions about the future fitness landscape. The predictions I dare to make are for trends I expect to be stable over a very long time. I will also assume that any advanced technology that we develop will spread to all members of humanity and rapidly so. Affluent and innovative countries will have early access. Still, I doubt that advanced medicine, say, will be kept a secret. New technology will spread from the origin countries to the rest of the world, and technological advances that have a significant influence on humanity will be available to all. Technology might be unevenly distributed over short periods, but I don't think that any subgroup of humanity can hoard a technology for long. Decades perhaps, but keeping a technology secret from others for centuries stretches the imagination, and for millennia it is inconceivable. Whenever I suggest that technology will fix or create a problem, I expect it to affect all humanity. Only with this assumption can we speculate on the evolutionary consequences of the entire species.

Future Medicine

It is always medical improvements that people bring up when they say that evolution no longer affects humans. It is undoubtedly true that modern medicine will strongly influence our future evolutionary development, but it cannot stop evolution entirely. If nothing else, we will always have a random frequency drift of neutral alleles, but I find it unlikely that selection won't be around as well. Medicines reduce selection against the conditions they mitigate or treat. Alleviating a condition lessens the selection against it, and curing the disease removes the selection entirely. Some congenital disabilities, that were lethal in the past, can now be cured medically or repaired surgically. Many conditions that would prevent a child from reaching reproductive age, or at least reduce its reproductive chances, are now curable. Of course, not all defects and diseases have a genetic component. Some are exclusively consequences of environmental factors, and some are poor luck. Such cases are not affected by evolutionary pressures, so we ignore them here. For conditions that have a genetic component, such medical advances have changed the fitness landscape.

A condition with a genetic component will have alleles that either increase or decrease the risk of said condition. Without effective medicine, we will see negative selection against alleles that increase the risk with a selection strength determined by the severity of the condition. Conditions that kill the affected before reproduction, that sterilizes, or dramatically reduce the chance of reproductive success, would have had strong selection against them, for obvious reasons. Many genetic diseases with less dramatic impact will have weaker selection against them.

Now, when modern medicine eliminates or reduces the severity of the disease, this selection will also be reduced or removed. It doesn't even have to be diseases that selection could be relaxed for. Minor inconveniences could be affected. Poor eyesight is genetic, and there is variation in the population, so clearly, we have the possibility of selecting for genes that remove the issue. The frequency of myopia is estimated to be around 22% worldwide, a high frequency that suggests that it cannot have been under strong negative selection in the past. Still, it is hard to imagine that there has been no selection. As a hunter-gatherer, poor eyesight must be worse than 20/20 vision; if you cannot see your prey, you are unlikely to catch it; if your search for food and your vision is impaired, you will not find as much. Today, however, myopia is no more than an occasional inconvenience. Glasses will fix the problem. (Glasses were invented around 1290, so this is not modern technology. On the time scale that evolution works on, though, 800 years is nothing). Contact lenses remove the issue as well, and if you suspect that glasses reduce attractiveness—which I find unlikely as a general rule—contact lenses will remove the selective disadvantage there is to that. We also have laser surgery, and I would be surprised if we will not have more options in the future. Any selection there might have existed against myopia is eliminated by technology.

So technology, especially medical technology, removes or reduces selection against traits that were previously kept in check by selection. In a game of 'survival of the fittest,' the less fit are no longer eliminated. The reason they are not expelled from the gene pool, of course, is that they are no longer unfit. The environment they live in now includes modern medicine. In this environment, the fitness landscape does not put their genotypes in a fitness valley. If we remove the technology again, the selective pressures would be back, and selection will work against the old diseases and defects. However, in the presence of medical technology, the affected are as fit as everyone else, and there is no selection against their genes.

Notice, however, that turning off selection against alleles that might harm us, if we are not treated with modern medicine, does not mean that these alleles will suddenly increase in frequency. Selection will no longer work actively to reduce such alleles' frequencies; when selection is turned off, their frequency will fluctuate randomly over the generations as neutral allele frequencies do. If an allele was kept at low frequency before, when it was selected against, it will still be at low frequency once the selection is turned off, and no mechanism will now force it to increase in frequency. It is more likely to be lost from the gene pool than to be fixed since it starts its random walk as a rare allele. Not that it matters; we have eliminated its harmful effect. We will not automatically get weaker as a species because we no longer select against these defects. We would see, however, that just by chance, some previously harmful alleles will rise in frequency or even get fixed in the population—alleles that could never have made it through the population when they were selected against. It will be very few—because only a few alleles manage to go from very low to a very high frequency without positive selection—but some likely will. Not that it matters; they no longer affect our fitness.

We could remove all medical drawbacks of a condition and still see a selection. If there are *cultural* reasons to avoid carriers of some allele or another or sufferers of one disease or another, then they will still be a selective disadvantage. Since 'survival of the fittest' really means 'reproduction of the fittest,' it is not only about alleviating defects and diseases. There could be social stigmas associated with conditions that are currently kept in check by selection. Today, HIV-infected people on proper mediation have the same life expectancy as non-infected, and they will not infect sexual partners. They have to take a regime of medicine every day but are otherwise living ordinary lives. Still, I think you will agree that many will hesitate with being in a relationship with an infected, all else being equal. While HIV infection is not a direct cause of a genotype, there is a genetic variation to how resistant you can be to an infection. With modern medicine, HIV will get you to reproductive age and beyond. Still, the stigma of being infected, if it continues into the far future, could give us a fitness gradient, and we might evolve stronger immunity to HIV. We could still see lingering selection after our science and technology have removed the cause of it.

In the next chapter, I will discuss the relationship between future medical technology and the changing demographics we are currently going through. For the remainder of this chapter, I will address two separate health-related issues that are likely to affect our genetic composition as a species as a consequence of modern and future living.

Infectious Diseases

Our greatest health risk is not faulty genes. We can be born with many defects, absolutely, but it pales compared to how many diseases we will be infected by throughout our lives. Viruses, bacteria, and other parasites, constantly invade our bodies, and infectious diseases kill over 17 million yearly according to WHO estimates. Yet, over the last few centuries, we have reduced the death rate to a fraction of the former worldwide level. It is not that fewer people die from infectious diseases now than in the past, but the individual risk has gone down. The population has grown at the same time as the death rate has decreased, and each year more people die from infections than were alive in the entire human species in the Paleolithic Stone Age. You cannot compare absolute numbers of deaths between now and the past, but if you correct for population size, the risk of dying from infection has plummeted.

Sanitation can take most of the credit. Clean water and

simple hygiene neutralize many of the infection pathways. Medicine obviously also contributes. Antibiotics and vaccines enormously alleviate the dangers of infection. Reducing the rate of infection and the severity of catching an infection has consequences for our disease resistance and will guide its evolution.

Through our evolutionary past, we have always had some resistance to infectious diseases. Our defense against infections is far from perfect, but we wouldn't survive without it. Whenever we have a genetic variation that increases or decreases susceptibility to a disease, there is a potential for adaptation. An infectious disease that very rarely spreads through a population does not affect fitness unless its mortality is incredibly high. It is here, and then it is gone, and adaptation does not have time to react. On the other hand, an endemic disease, one we are regularly exposed to, likely creates a fitness gradient, and the severity of the disease will affect the steepness of the slope in the fitness landscape. If we replace natural resistance to a disease with technological protection—vaccines, cures, or perhaps eradication of the disease—then we have an arms race between technology and infections but not an evolutionary arms race. We will effectively have turned off natural selection against that disease. You could say that modern medicines turn off evolution—a claim made by many—but it is only valid in this specific area. If our sole defense against diseases is technological, then there are no fitness gradients for evolution to work with. We won't improve our fitness, but neither will we grow less fit—there aren't any hills and valleys in the fitness landscape any longer.

That medicine will reduce the power of selection is not the only possibility. Sometimes, removing one selective pressure creates another. Consider a particularly interesting genetics/disease case. There is a gene that causes the disease called sickle cell anemia. The condition causes the red blood cells to change shape—into a sickle-like shape, thus the name—with various detrimental consequences. There is a 'healthy' and a 'disease' variant of the gene, and to get the disease, both of your two gene copies must be the disease variant. If you have one or two healthy copies, you are fine. Selection is weak against such gene variants when they are at low frequency because then we rarely see two of the harmful variants in the same individual, and there is no fitness disadvantage to carrying only one. At higher frequencies, we start to see individuals with two deleterious copies. These people have reduced fitness, so now we see selection against the disease variant. Selection will push the frequency down. Any allele that is kept at low frequency will eventually be lost, so selection will get rid of the disease-causing variant eventually. For this particular gene, however, there is also a positive selective pressure when the disease-causing variant is at low frequency. It just so happens that those that carry one (but not two) copies of the disease variants have an increased resistance to malaria. At low frequencies, we practically never see the sickle cell consequences of the gene, but we do observe the malaria resistance. Thus, the disease gene goes up in frequency. At higher frequencies, we see the sickle cell effect, and the frequency goes down again. These two forces balance each other, so the allele stays at a moderate frequency. If we remove the selection in one of the directions, then we should see the fixation of one of the two variants. If we have an efficient vaccine against malaria or a cheap and readily available cure, then we would no longer see a positive selection for the sickle cell allele at low frequency. We would still see a negative selection at higher frequencies, which would push the frequency of the disease allele down, and it would be purged from the population.

I find this example interesting precisely because it is not just a case of relaxing selection for resistance to an infectious disease when we have an effective medicine. Instead, it is selection against a genetic disease as a consequence of better medicine against an infectious disease. Medicine changes the selective landscape, but in this example, it doesn't eliminate selection, it merely alters the direction.

Medicines do not just affect our own evolution (potentially by reducing selection); it also forces the targets of the medication to evolve. This is why, for example, resistant bacteria is a problem. When we develop a new drug, we put extreme selective pressure on the organism it is aimed at, and an intense selective pressure means swift evolution. If there is a small fraction of a bacterial strain with some resistance, they might initially be at a low frequency because they are less fit than the rest of the bacterial population. However, if we subject the bacteria to antibiotics, all the non-resistant bacteria die, and we are left with a population where all are resistant. Now, the resistant variants are not kept at low frequency by the higher fitness of the other bacteria, and they can enter an explosive growth. This effect, combined with the short generation times of many bacteria, means that resistant strains can evolve exceptionally fast. Penicillin was discovered by Alexander Fleming in 1928 but it wasn't widely used as a medicine until 1940. As soon as 1950 we saw the first resistant strain, Staphylococcus aureus. This strain has now evolved into MRSA, a multi-resistant variant, i.e., one that is resistant to more than one antibiotic. The more antibiotics a bacterium is resistant to, the harder it is to deal with it. Bacteria can often exchange genes horizontally, meaning that they can swap genes with other bacteria and not only pass them on to offspring. Horizontal gene transfer makes resistant bacteria even more dangerous. If a bacterium develops resistance and passes it on to other bacteria, then the resistance can spread through the population at a much faster rate than would otherwise be possible. They do not need to wait for many generations before a resistant allele is fixed.

We are, of course, developing new antibiotics. If a bacterium is resistant to one antibiotic, we can kill it with another. But there is an arms race going on. We develop new antibiotics, and the bacteria get multi-resistant. We get a new antibiotic to deal with the resistant strains, and they evolve resistance to that antibiotics as well. There is no guarantee that we will win this race. If more bacteria develop multi-resistance, so all our antibiotics are useless against them, and if they then transfer those genes horizontally, we are looking at a health disaster. Before we had antibiotics, an infection that is harmless today could kill you. The thought of going back to how it was before we had antibiotics is frightening.

If the effectiveness of our antibiotics is reduced substantially in the near future, it would not drastically affect our genetics. Many will die that we could cure today, but since medicines like antibiotics have been with us less than a century—and that is only for the affluent countries—it has not changed the genetic composition of humankind yet. If our drugs lose their power tomorrow, then we would go back to the selective pressures we had before we got them, and we would continue along the evolutionary trajectory that our medicine suspended. However, if we have the protection of modern medicine far into the future, and it loses its effects, then we might return to an ancient selective pressure with a weakened immune system.

If our drugs turn off selection for a robust biological immune system, this system will weaken. Over a sufficiently long time, say up to a million years, mutations could destroy some of the genes that used to protect us from pathogens, and the defective alleles could spread throughout the population. They are unlikely to do this to a large degree, though, because of their initial population frequency. To destroy a gene, the protective alleles would have to drift from high to low frequency and the defective alleles from low to high frequency. As we saw in the previous chapter, this is unlikely to happen by random drift, and if it happens, it takes an exceedingly long time.

Yet, an entirely technological defense against infections over a sufficiently long time could theoretically weaken our immune system. It cannot happen if our medicines lose their efficacy in mere centuries from now. It will not occur in a few millennia, although our immune system could be weaker if it hasn't been utilized for thousands of years. After a million years, some protective genes could get lost because of relaxed selection, and the consequence of this, if we are once again the targets of pathogens, could be disastrous. It is inconceivable that we could lose our immune system entirely. We will never live in an environment where there is zero chance of infection; microorganisms are everywhere, so the immune system will always be exercised. But we could end up with such a robust artificial defense against germs that our natural defenses atrophy. As long as our synthetic protection is working, this does not impact our fitness. Still, it would leave us immensely vulnerable to any microscopic enemy that makes it through our technological defenses.

Still, for the moment, improved medicine—new vaccines and new cures—has lessened the risk that infections pose. But social changes work against this and increase the infection rate of contagious diseases, and any of the two forces—improved medicines or increased infection rates—could win. The primary social changes that will increase the spread of diseases are denser populations and increased mobility.

Diseases can transmit person to person, like the flu, measles, AIDS or COVID-19, or through an animal, like Zika, typhus, malaria, and the plague. Consider the first case. The number of susceptible people each infected person comes into contact with determines the rate with which the disease spreads. Urbanization increases the population density, the average infected person comes into contact with more susceptible people, and the infection spreads faster. As the population density grows, the speed of infection will increase, all else being equal, and the urbanization trend we see today will pack us denser in the future.

The case where a disease does not spread directly between people, but through one or more animal hosts, is somewhat different. The infection rate does not depend on interactions between susceptible and infected people, but between susceptible and the animal hosts. But population growth can still have an effect. In overcrowded living conditions, people often get into closer contact with animals, both pests and farm animals, and this can increase the infection risk. This is mostly a problem in developing countries or slum settlements, and I imagine that it will be less of a problem in the future, depending on whether population density or wealth grows faster. But the risk of infections from animals can also increase as a consequence of population growth in another way: by humans expanding into the animals' habitats. As we expand into animals' habitats and destroy them, as we are wont to do, we force the animals into either extinction or into adapting to *our* habitat, cities. Many animals have already adapted to urban life, raccoons, pigeons, rabbits, foxes, crows, , etc. If we force more animals into urban life and thus increase our contact with them, we could very well run into a new disease that way.

The higher mobility of the future, e.g., cheaper and more frequent plane travel, might also increase the speed at which a disease can spread. We can propagate a disease that infects through an animal host by spreading the host. Likely, trading ships were part of spreading the rats that gave us the plague. Higher mobility, e.g., more trading lanes, can help spread an animal-borne disease. However, the worst-case scenario for spreading a disease worldwide is the person-to-person transmission, so let us consider that. The number of people each person can infect depends on the fraction of the population that is susceptible; immune or dead people do not get infected, and the infected are already infected. If everybody that an infected interacts with is vulnerable, the infection rate can be high, and this is the situation in the initial phase of an epidemic. As the disease spreads, however, a more substantial fraction of the population will be infected, and of those, some will recover and be immune, and some will die. An infected can only spread the disease to a susceptible. As the fraction of the population that is susceptible goes down, the average number of people that an infected can pass the disease to go down as well. If there are no vulnerable people left, the disease dies. Often, though, the infection rate merely slows down, and we have an endemic disease that still infects new susceptible but does not spread aggressively through the population. This can happen when a sufficiently large proportion of the population has developed some degree of immunity.

A disease that spreads through a single geographic area will see the ratio of susceptible to infected people go down, and the infection slows down as well. If an infected now travels to an area where there are no infected or immune, but only susceptible, then the disease can, once again, spread aggressively, an effect known as a *virgin soil epidemic*. An extreme example from history is when Europeans arrived in the Americas. The Europeans had developed some immunity to a range of diseases that were unknown in the Americas, including cholera, smallpox, measles, and many more. The American populations had never been exposed to them and had no immunity. We do not know how many were killed by these, to the native populations novel, diseases, but estimates go as high as 90% of the population.

Admittedly, this is an extreme example and not something we will experience again. We no longer have large populations where one can develop immunity isolated from another.[1] Still, a disease can arise that quickly spreads through a local community. Many might get sick and die, but if no one travels, the disease will burn through the

---

[1] There are isolated tribes in pockets around the world, but when considering human evolution, we can ignore them.

population, and the epidemic will die out. If people move to neighbor communities, the epidemic will flare up again there. It would still be localized, though, and we could fight the disease with quarantines. With global travel, infected people could spread the disease to all corners of the globe, even before we discover that an epidemic is rising, and we quickly have a global pandemic.

A related point to add to mobility is incubation and recovery time. The time it takes for an infected to show symptoms (if he does) and until he recovers, is relevant for how he can spread the disease. If a pandemic is raging and governments quarantine people with symptoms, infected will be prevented from traveling once they show symptoms. If travel time is slow, they will be put into quarantine before they have traveled far. If travel time is fast, they can spread the disease over many places before they are stopped. Similarly, if you can travel far before you recover (or die), you can infect more areas than if you can only travel short distances. The easier global travel is, the faster we can spread a disease from its origin to the entire world.

Throughout history, we have repeatedly been hit by new diseases we have had little defense against, and this will happen again. Diseases never appear out of nowhere, so whenever we talk of a new disease, we only mean a disease we haven't encountered before. Usually, this is a disease that jumped to humans from some other animal, called zoonoses, or it is a mutant of an existing human disease. Many diseases mutate into new forms. Influenza, for example, is endemic, but it mutates sufficiently fast that we do not grow immune to it. Sometimes the mutations change the virus enough to make influenza a pandemic that kills tens or hundreds of millions. The Spanish flu,[2] an H1N1 influenza, is estimated to have killed 50–100 million people. Influenza is also the poster child for a disease that jumps species. The bird flu is an influenza that is endemic in birds that jumps to humans. Another example of a virus jumping to humans is SARS-Cov-2, the virus that causes COVID-19. It likely originated in bats but jumped to humans and quickly spread to become a global pandemic with disastrous results. Regardless of where a new disease comes from, we will usually have no immunity, which can make the mortality of the disease frighteningly high. Luckily, in the cases of diseases jumping from animals, the disease is no more adapted to us than we are to it. Usually, this means that it does not infect us efficiently or transmit from human to human, which slows it down and gives us time to react. We are not always this lucky, of course. The scary scenario is when we see a disease appear with high mortality and high infection rate, like mutated influenza. There are warning signs before we have a pandemic, it starts as a localized epidemic, but the time we have to react is short in an age of dense urban areas and global travel. What begins as a minor epidemic in one corner of the world can be a worldwide pandemic three months later. We can attempt to slow the disease down by identifying and isolating the infected or take social measures to limit interpersonal interaction, but with our current technology, we cannot develop vaccines and medicines fast enough to counter the speed at which a disease can spread. As

---

[2] The Spanish Flu did not originate in Spain, as the name might otherwise suggest. Its origin was in all likelihood Kansas, USA. The US press was under censorship because of the first world war, while Spain's was not (Spain was neutral in the war), so Spanish newspapers were the first to report on the pandemic.

our medical technology improves, we will improve on our reaction time, but whether this can compensate for our social transition to a more connected world is not clear.

There will, of course, also be a delay between the appearance of a new disease and an evolutionary adaptation to defend against it. But unlike most drugs, our immune system is a universal defense against infections. It is trained to protect us against a wide range of infections, including pathogens it has never seen before. If the future is one of many more infections, then there would be an increased selective advantage to a stronger and versatile immune system. If our technology cannot compensate for the increased population density and increased mobility, then we could be looking at a future with stronger and not weaker immune systems.

All pandemics are horrible, with many tragic losses of life, but they are not all equal; some are truly devastating. Consider just influenza. It is usually not a concern, but it can turn very deadly. In the 20th century, we had three influenza pandemics, the Asian Flu (1957–1958), with 2 million deaths, the Hong Kong Flu (1968–1969), with 1 million deaths, and the above-mentioned Spanish Flu (1918–1920) with up to 100 million deaths. In the 21th century, the swine flu killed up to half a million people. Two to three orders of magnitude in fatalities are a dramatic difference, and the Spanish Flu is one of the deadliest in history. If the Spanish Flu killed 100 million—the global estimates are somewhat uncertain—it would be around 5% of the worldwide population (a global population with just short of two billion people in 1918).

I am mentioning the flu because few realize just how dangerous it is. It is, of course, not the only pathogen that flared into pandemics in the 20th and 21st centuries. In the 20th century, cholera killed more than 800,000, sleeping sickness more than 1.5 million, smallpox more than 300 million, and AIDS more than 32 million. In the 21st century, the largest pandemic is COVID-19. At the time I am writing this, more than a hundred thousand have died, and the death count is still rising at an alarming rate. Although the death count for smallpox is three times higher than for the Spanish Flu (and much higher if we include previous centuries), they were stretched over many more years. In the last 100 years before smallpox was eradicated (last occurrence in 1977), the virus killed 500 million, or an average of 5 million per year. The Spanish Flu killed 100 million over two years. We have lived with smallpox since prehistory, but the Spanish Flu came unexpectedly out of nowhere. Smallpox killed more; the Spanish Flu was more dramatic and shocking to the world.

This death toll of the Spanish Flu is comparable to the most famous disease in history, the Black Death (1331–1353), that also killed around 100 million (75–200 million). The Spanish Flu is comparable to the Black Death in absolute death count, but of course, the global population size is not comparable between the 14th and 20th centuries. Estimates put the global population size at less than half a billion in 1331. In Europe alone, The Black Death killed 10–60% of the population.

As far as we know, The Black Death predominantly spread through fleas and rats and not direct transmission between humans. Today's hygiene in the developed world could largely have prevented the spread, and with antibiotics, we can treat it if we act fast (Very fast! Hours, not days). The

Black Death was probably a mix of three plagues, predominately the bubonic plague, with a mortality rate around 30–90%, the pneumonic plague, that can also transmit directly between humans, with a mortality rate of 90–95%, and the septicemic plague with mortality near 100%.[3] All three plagues are infections by the bacterium *Yersinia pestis*, and they differ in where the infection is, lymph nodes, lungs, and blood, respectively. The plague is still around today, there are occasional outbreaks, but with infection numbers in the tens or hundreds. The plague can no longer be considered a global threat. Influenza, obviously, is also around and kills around 250,000–500,000 a year. We see influenza pandemics every few decades, and while most are relatively mild, we know that the flu can turn incredibly deadly, so this is a concern. You probably do not worry much about the seasonal flu, but a pandemic that kills one out of twenty in two years, as the Spanish Flu did, should concern you. A pandemic that potentially kills half the population, as the Black Death might have done, should terrify you.

It is horrifying to think that another disease, as virulent and infectious as the plague, could rapidly spread through the entire species and potentially kill half the human population. Or 75% of the population. Or 99%. We might not be this vulnerable. We might never see such a deadly disease, and if it appears, we might be better equipped to deal with it than our ancestors were. We have hospitals to alleviate symptoms, and we know how to quarantine and

---

[3]The mortality rate is not the fraction of a population that dies of the disease. It is the number of infected that does. A disease with a high mortality rate that infects few people will not be as severe as a disease that infects many people but has a slightly lower mortality rate.

isolate infected, although our track record in doing this is not stellar. We might get lucky and avoid such a terrible scenario, but what if it happened? Even if nature doesn't send such a disease our way, we could do it to ourselves. We already have the necessary technology to create a synthetic virus. That technology will be widely available in the future. Some rogue agent could manufacture and release a truly devastating pandemic, an almost existential thread.

If a significant fraction of humanity perishes, then the genetic composition of our species will shift. Allele frequencies can move considerably faster in such an event than they can in a large steady sized population. In a smaller population, the random drift of frequencies is faster than in a large population. If a pandemic reduces our population size, allele frequencies will consequently change faster. Even if the reduction in population size caused by a pandemic is short-lived, and we bounce back to billions of people after that, the population bottleneck that we would go through would speed up allele frequency drift. Once we are out of the bottleneck, we have a slow genetic drift again, but with a new allele composition.

Speeding up drift will change our allele frequencies randomly, but if there is genetic variation in the resistance to the pandemic, then selection for that is dramatic. Even in a small population, where selection is generally less effective, we will see resistance alleles shoot up towards frequency one. The selection strength in this scenario is extreme, and this will more than compensate for the reduced population size. It would not be unlike what happens with bacteria exposed to antibiotics; the resistance alleles might be at low frequency before the event, but the frequency blasts up and is high after the event.

It is undeniable that vaccines, improved hygiene, and modern medicine have reduced the threat of infectious disease and saved countless lives, but they might also have had negative consequences. There is some evidence that improved hygiene increases the risk of allergies and autoimmune diseases. The so-called *hygiene hypothesis* states that infections early in life, especially infections by microbes that co-evolved with humans and that we are resistant to, is necessary to train our immune system. Put simply; the immune system needs to learn to distinguish harmless from harmful to develop correctly. Otherwise, if you are exposed to harmless molecules such as traces of peanuts, the immune system will consider it dangerous and shift into overdrive and start hurting your own body. Many of the autoimmune diseases and allergies that are believed to be caused by better hygiene only rose in frequency in the last one or two centuries, so we have not had to evolve resistance to the problem. Now that the problem is here, we have a selective gradient that can drive evolution. We cannot merely lower our hygiene. Fewer infections are still preferable to the risks incurred by good hygiene—we do not want to go back to the high child mortality of the past. So training our immune system to be less aggressive, teaching it not to overreact to harmless exposure, adapting it to an environment with fewer infections, would be highly beneficial. It doesn't have to make our immune system stronger or weaker, necessarily, just better at distinguishing good from bad without the training it evolved to need.

However, the simplest way to reduce autoimmune diseases would probably be to reduce the effectiveness of the immune system. Selection prevents this, there is a considerable selective pressure to keep it strong, but if we reduce the need for our built-in immune system, a selection against autoimmune issues could arise. If we remove the need for a robust immune system, it could weaken over a million years, but it would do so via random processes, and this would be extremely slow. If there is an active selection to weaken the immune system, then the process would be accelerated. Reducing the selection in one direction could create a selection in another direction, similarly to how a malaria vaccine could introduce selection against sickle cell anemia. We could see improvements to medicine and hygiene reduce selection for a strong immune system and increase selection against autoimmune diseases. A long period with an effective artificial defense against infectious diseases could radically change our inborn defenses if there is a selection for doing this. With disastrous results, if we suddenly need it, if a pandemic arises that our technology is not equipped to handle.

In a future with better hygiene, vaccines from more diseases, and improved medicine to cure infections—a technological rather than innate protection against pathogens—we might see a degrading immune system. We will not have the selective pressure to keep the immune system strong, so random processes—mutations and genetic drift—could erode it. If there is selection on top of the random processes, because our immune system is a potential danger to ourselves, then the erosion can happen quickly. Working against this is the higher infection rate that higher mobility and increasing population densities will likely cause. So arguments could be made for both weaker and stronger immune systems in the future. I do not doubt that our interaction with infectious diseases will change in the future, or that this change will affect us genetically, but

I am not brave enough to wager in what way.

The New Killers

In the developed world, the leading killer is no longer infections. It used to be, but now the two foremost causes of death are cancer and heart disease. As we get better at avoiding and curing infections, the percentage of other conditions must go up, of course, because we still die.

Let us first consider cancer. We have always had cancer, but our environment strongly affects our risk, and our environment has changed dramatically over the lifetime of our species. Pollution, for example, is a major cause of various cancers. Pollution isn't a new phenomenon, though. When we hear pollution, we might think of smog in inner cities, but a wood stove also pollutes. Smoke from a wood stove increases your risk of cancer many times over that of exhaust from a diesel truck. A gas stove is better than a wood stove and an electric stove even better, and the typical household in the developed world now uses an electric stove. If we consider only cooking, we are reducing our risk of cancer. But there are, of course, other sources of carcinogenic pollution in modern urban life, and whether this has increased or decreased the risk of cancer is hard to tell. A CDC report from 2017 showed that the rate of various cancers in the US was higher in rural areas compared to urban areas. Cities in the 21st century are substantially less polluted than in the 20th century and earlier, and that might have something to do with the reduced risk in urban areas. Perhaps city pollution today isn't increasing the risk as much as one might think, relative to pollution outside cities. Still, since modern rural life is very different from life in the Stone Age, we cannot conclude from this observation that the cancer risk has been constant over time.

Other environmental factors have also changed since the Stone Age. Many of us spend more time indoors rather than in the sun than our ancestors did. Sunlight is a factor for melanoma. Pollution can have increased our risk of cancer, but less exposure to sunlight could have decreased it. We have changed our diet, and if you follow the news, you will know that practically anything you eat or drink affects the risk of cancer in some study or other. Processed food, which they didn't have in the Stone Age, increases the risk of cancer. Food prepared over a fire, which Stone Age humans had more than we do, also increases the risk of cancer. We don't have cancer studies from the Stone Age, or even a few centuries ago, so we don't know if our environmental risk is higher or lower than theirs. It could be a wash.

We may or may not be at a higher risk now than earlier in our species' history. The only reason that cancer is high on the list of causes of death is that other causes, that medical technology has dealt with, no longer have prominent places on that list. The environmental factor for cancer have changed over time, but maybe the risk hasn't. If we pollute less in the future, our environmental risk could go further down, and, consequently, so should our overall risk.

For selection, however, it doesn't matter what some environmental risk is, what matters is the risk that someone dies of cancer, and that risk has gone up as other risks have gone down. If we pollute less in the future, which I think we will, and if we get better at treating cancer, which I also believe we will, then the fraction of deaths attributed to cancer will go down. I don't think we will see a dramatic drop in cancer deaths, however. The medical improvements that eliminated other disease risks help us live longer, and longevity is also a variable for cancer. Cancer is just a waiting game—eventually, there will be failures in our cells that trigger cancer, so if we wait long enough, cancer is unavoidable. We will get better at treating cancer, and yet cancer will pop up more often, percentage-wise, as we live longer, precisely because we live longer.

We know of many alleles that increase the risk of various cancers, so there is clearly a genetic component to how resistant we are to cancer. However, if cancer predominantly causes death late in life, after reproduction, you could argue that it doesn't have much of a fitness effect. Still, a little selective advantage suffices for changing our genetic composition, and not all cancer strikes at middle-age or older. There has presumably always been some selection against cancer, even when it wasn't a leading cause of death. Our genetics doesn't care if the risk posed by smallpox, for example, is greater than the risk of cancer; if it sees a selective advantage to avoiding death, it will go for it. The *strength* of the selection, however, depends on how many individuals see the effect of increased resistance to cancer. That, in turn, depends on how many people live to the age where the risk is appreciable. People live longer now, consequently, more get cancer, and that strengthens the selection against cancer. As for the argument that this selection will be weak anyway because cancer usually kills after reproductive age, this probably isn't true. Even when cancer doesn't strike until after reproduction, there can still be a selective benefit to cancer-resistant genes. We should not under-appreciate the help grandparents can provide in raising and supporting grandchildren. We not only live longer now, but we are also active at a more advanced age than before, so grandparents can help grandchildren or great-grandchildren for longer. If the support of multiple generations increases the fitness of a child, then we see a selection for living longer, including a selection against cancer.

Since more people are at risk of dying of cancer today, and since there is an adaptive benefit to living long past reproductive age, we can see a strong selection against cancer in the future. We won't eliminate cancer. Cancer is a failing of our own cells, and our cells are continuously damaged by environmental factors and simple wear and tear. This damage is so prevalent that we all have cancer cells in our bodies right now; they just don't multiply since our body is good at getting rid of them. Our defenses can't be successful forever, though, and eventually, some cancer cells will multiply and spread. The longer we live, the higher the risk that it will happen to us. Still, if we select for resistance to cancer, we can delay the time until cancer threatens us, and maybe, if we delay cancer enough, something else will kill us first.

The greatest threat to public health in affluent societies is likely obesity. Obesity is a concern since it is the cause of many health risks, including cancer, diabetes, and cardiovascular disease. Cardiovascular diseases kill twice as many as cancers each year, ~17 million, and are the worldwide leading cause of death (cancers only kill ~9 million per year for comparison). The primary cause of obesity is that we consume more calories than we burn. It sounds like a case of overeating, and to some extent, it is, but it is more complicated than that. No one can gain weight by consuming fewer calories than they need, but the calories we burn depends on multiple other factors, including genetic factors. These factors, combined with the vast selection of calories readily available to modern humans, is what leads to the obesity epidemic.

Obesity is something that will almost certainly affect our future genetic composition. We have not been subject to obesity in the deep past. There are two reasons for this: in the past, we had more manual labor that helps us burn calories, and we had less available high-calorie food. Because obesity is a new phenomenon, we have not had time to adapt to the problems it incurs. We started having readily available calories shortly after the invention of agriculture, but only recently have we had high-calorie food in abundance everywhere. There is, of course, still hunger in the world, but the fraction of humanity at risk of starving is plummeting, and in the future, I would be shocked if starvation is still an issue. Obesity probably will be.

The evolutionary issue with obesity is more than a lack of selection against it. We presumably selected for genes *for* storing fat in the past, which is the reason that obesity is a global problem today. We adapted to store fat whenever we could. Throughout our species existence, available food was scarce and unpredictable, and we needed a strategy for surviving between periods of available food. As a consequence, we have evolved bodies that will store fat as much as they can, and we have developed brains that prefer high-calorie food to give us that fat. We like the taste of fat, salt, and sugar because we evolved to like it. It gives us what we need to get from one meal to the next, in a world where meals can be far apart. It is not an ideal strategy when meals are plenty and frequent. We are in a situation where traits that we selected *for* in our past, because they were good for us then, are now traits that are harmful to us, and we need to change the direction of the selection.

So how could we imagine to evolve now, to save us from obesity? One noticeable trait could be a higher metabolism. If we burn calories faster, we do not get obese. Some people can eat much more than others and still stay thin. Their metabolism burns away the extra calories. If we have a substantial selection gradient for those genes, leading us away from the obesity valley in the fitness landscape, then we should soon have a much higher metabolism than we have today.

Speaking of burning more calories, we know that some people are more likely to exercise than others. Maybe through greater willpower or simply because they like exercising more. Others just prefer to watch Netflix. If the genes for 'exercising' spreads through the population, this could be a step against obesity. If it is 'greater willpower' genes, it could affect much more. Imagine if, to avoid the risk of obesity, we select for genes that increase our resolve and drive in general. That could change our society profoundly. There is likely already selection for such traits, but if they tag unto selection against obesity, they become life or death traits. In that case, the selection could well be incredibly rapid and could drive our psychological evolution as well.

Burning more calories is not the only solution, of course. We can eat less. Maybe we can decrease our appetite. The main reason smokers gain weight when they stop smoking is that smoking suppresses hunger. If we could get that hunger-suppressing effect without tobacco, that could be good for us. We could also imagine that our preferred food changes. We evolved to prefer high-calorie food, but we could change that and evolve to prefer a salad. If we prefer salad over burgers, we will have alleviated much of the obesity problem.

If we give it a million years, there is a chance that we could eliminate the risk of obesity through selection. There are already alleles in the population that can do this, e.g., a higher metabolism or a desire for exercise, and they just need to rise in frequency. If obesity remains a high-risk phenotype, there will be a selection for these traits. If the beauty ideal is slim rather than fat in the future, there will also be a sexual selection. The same attributes that protect you from obesity make it easier for you to attract mates. These two forces, acting together, can drive an extremely rapid adaptation.

I am less optimistic about selection against cancer. It is evolutionarily possible to move the cancer risk close to zero. Naked mole rats, underground social rodents from East Africa, do not get cancer. We don't know exactly how they manage this, although there are several theories. Regardless of how they do it, it might not be possible for us to evolve the same cancer defenses. It may be as hard as evolving an octopus eye. For us, we know many genes linked to cancer, but the alleles we know about, with few exceptions, only influence cancer risk by tiny amounts. Combined, they can have a substantial effect, but individually they do not, and selection works on individual alleles. There will be a selection against cancer, I am convinced of that, but it could work extremely slow, and likely there is a limit to how low the cancer risk can go in a million years. My money would be on a technological defense from cancer, much before we evolve into a cancer-free species.

Regardless of how we do it, evolutionary or medical, if we manage to bring down the risk of dying from the current major threats, new ones will appear. Until we achieve immortality, death must have a cause, and death statistics are only giving us the frequencies of deaths from various causes. Every time one risk goes down, others must necessarily rise. There will always be new health risks that we must either adapt to or develop medical treatments for. I believe that we are well equipped for both in our future; our technology improves at an exponential rate, and our large population size has increased the efficacy of selection.

CHAPTER 4

Demographics

Infant and child mortality was incredibly high in the past. It still is in some parts of the world today, but the mortality decreases with better sanitation, higher hygiene, and better medicine. In 1800, even in the riches countries in the world, 30% of children died before the age of five. In poorer countries, this was as high as 50%. In the 1950s, the worldwide child mortality was just short of 20%—that is one in five children. That was the global average; rich countries were naturally better off with numbers in the single digits. Today, the rate is around 5% worldwide—in developed countries, about 0.5%. The numbers are still dropping in both rich and developing countries.

The drop in child mortality over the last few centuries is astonishing, and because it has dropped, life expectancy has shot up. The global life expectancy has gone up from 30 years in the pre-modern world to over 70 years today. There is a vast disparity in life expectancy between rich and developing countries, but it goes up everywhere. No country today has a lower life expectancy than the countries with the highest life expectancy two hundred years ago. Life expectancy is rising fastest in developing countries where child mortality is still high, but now rapidly decreasing. If you decrease child mortality slightly, you increase the mean lifetime substantially. But it is not only the child mortality that drives up our life expectancy. At all ages, we see an increase in the expected remaining lifetime, the so-called *survival rate*. Someone at age 50 today can expect to live longer than a 50-year-old a hundred years ago—the same for someone aged 70 or 90. The survival rate doesn't grow evenly at all ages, of course. When life expectancy rose from 30 to 70 years, it was for newborns. It didn't increase by 40 years as well for those that were already 50. The older you are, the less you have gained in your remaining life expectancy, but at every age group, you can expect to live longer today than you could in the past.

A society with high child mortality tends to have many children. If most children die, you need to have many of them so that some survive and preserve the population. In such a society, there will be extremely many children, but as you look at older and older ages, the number of people drops rapidly. For every child born, maybe one in five makes it to age 10. For every five children at age 10, perhaps just one makes it to 20. If so, the ratio of newborns to twenty-year-olds is 25; you need 25 infants to make one 20 years old. If we take such a society and lower child mortality, i.e., we do what we have seen happen in history, then what will happen? There will still be many children, but now they survive to have children themselves, and their children do the same, and so on. We no longer see the sharp drop-off as age increases—people survive longer—so the ratio of young to old is less extreme—but still extreme. The population grows at an exponential rate, which the demographics will reflect. Let's say that each couple has four children, and let's say that all four children survive and reproduce, where perhaps earlier, only two survived. Having four children is low for such a society, but it makes the following math easier. So, each couple has four children, which means two couples have eight. The eight forms four couples, and each of those has four children, so 16 children or eight couples. The eight couples have 32 children, which gives us 16 couples, and these 16 couples will provide us with 32 couples that will, in turn, give us 64. So we see the growth: 2, 4, 8, 16, 32, 64, ... Each generation is twice the size of the previous.

Exponential population growth is not sustainable. Such a society will quickly run out of food and other resources. This should worry us, since the global population size *is* growing exponentially, and it is predominantly driven by populations going through this phase of their development. But it is only a phase. Societies do not continue exponential growth for many generations, and the growth usually stops before they run out of resources. They do sometimes run out of resources, there are famine and overexploitation of nature, but with few exceptions, societies change before we see a population collapse.

After the child mortality drops, with some delay, birth rates go down as well. Societies go from a population in equilibrium with high birth rates and high mortality rates to one with high birth rates and low mortality rates, with explosive population growth. But then they change into a society with low birth rates and low mortality rates, where the population size is stable once more, although substantially larger than it was before it all began. This process, from a stable population with a high mortality rate, through an exponentially growing population, and finishing in a stable population with a low mortality rate, is called *the demographic transition*.

Although I say that a low birth rate follows a low mortality rate, it is a correlation and not necessarily a causal link. There are other explanatory variables. The lower mortality rate usually correlates with increased wealth and higher education of women as well, and it is more likely that those two cause the birth rate to go down. For our discussion here, however, it isn't crucial through which mechanisms we transition through the three different stages, only that we seem to do.

The demographic transition isn't happening concurrently on a global scale. It has completed many places, e.g., Europe, China, Japan, and North America. Many African countries are still in the middle stage, with exponential growth. Few societies remain in the initial stage. Sadly, many still live without access to clean drinking water and effective medicine. Nevertheless, generally, health is improving, child mortality is going down, and the few remaining populations are beginning to transition. The transition speed varies as well as the timing. Western Europe went through the transition first in a period of 2–3 centuries, from the 18th to the 20th century. Other countries went through the entire change within the 20th century. It took 95 years for the average woman in the UK to go from having more than six to fewer than three children. It only took 10 for the average woman in Iran. Some countries seem stuck in the middle phase at the moment, but I expect this is temporary and that they will continue the transformation soon. Countries that are starting the transition now can use the experience and knowledge of those that have already completed it. Sanitation and medical knowledge can move them from the initial stage to the second stage, and perhaps they can find inspiration from those in the final stage for how to get there as well.

The global explosive population growth is driven by those in the middle phase of the demographic transition, and consequently, we have global exponential growth. The rate of the population growth is decreasing, however, as more populations enter the third stage. If the entire human population moves to the final stage of the transition, and I think they will, then the population size will even out. There will be a lag from when all reach the final stage, and the population growth hits zero, since younger generations will still be larger than the older generations for a while, but eventually, the increase should stop. At the end of the transition, we will have stabilized with a new demographic than what we evolved with. We will have a society where the fraction of older people is much higher than it ever was in the past. This will also shift the life stages that our genes are exposed to; in the past, most of the time, genes found themselves in young people, and selective factors relevant in youth drove our evolution. Now and in the future, they will be exposed to later stages of life, and we could see a selection for traits that are important there. In this chapter, I will discuss the evolutionary consequences of the demographic transition.

Birth Rates

It is an accepted fact in social science that as wealth and education go up, the birth rate goes down, and this is what we observe in the demographic transition. But is there any evolutionary reason to reduce our birth rate? That might be a little more complicated.

If we zoom out from humanity and consider living organisms in general, there are two extremes in reproduction strategies, called r and K. Those names are not easy to remember, so I will call them 'many' and 'few.' The 'many' strategy is, as the name suggests, to have many offspring. Turtles, for example, lay many eggs as do spiders and mosquitoes, rodents have large and frequent litters; they follow the 'many'-strategy. It is an excellent strategy if most of your offspring die early, because, by having many, some might survive and spread your genes. It is also an excellent strategy in an environment where there is room to grow and plenty of resources. If the population can grow exponentially, then the higher the birth rate, the faster you can colonize your environment. And if you reproduce more rapidly than your competitors, it will be your genes that take over the future species. The strategy is damaging for you and your offspring if the environment cannot support the exponential growth. The population will soon exhaust all its resources, and the population collapses. The strategy makes it easy to bounce back again, once the resources are replenished, but it is wasteful to produce many offspring only to see them compete against each other until the population collapses again. The 'few' strategy is to have fewer offspring, but protect and nurse of them, so they survive to reproduce themselves. Many mammals and birds take this approach. The 'few' strategy works when there are limited resources in your environment. You don't produce so many offspring that they exhaust the available resources, and because you protect your offspring, the few that you do have will survive to reproduce. It is an excellent strategy if you live in a stable environment with limited resources.

Humans are firmly in the 'few' category. Yes, until recently, child mortality was high, and we had more children as a consequence of this, but we still had very few compared to most other animals. We have to invest many resources on our children and care for them for years, or we couldn't be human. I do not merely mean that it would be inhuman not to care for our children, I mean that we *couldn't* be the species we are if we didn't. The human brain that gives us our immense selective advantage over most life on earth is too large for any strategy that doesn't involve prolonged and intensive care of offspring. The brain cannot fully develop in the womb; the baby's head would have to be larger at birth if it did, and it is already at the limit of what the human birth canal and pelvis can handle. We are born underdeveloped, a human baby is entirely helpless, and in the first handful of years, we would never be able to survive on our own.

We are a 'few' species, and I didn't bring up the two strategies to talk about 'many' species but to bring the two strategies into a discussion about human reproduction. I am not, in any way, implying that we are near a 'many' species— those species produce vastly more offspring than we ever will. But I will use the terminology from ecological theory. We have one strategy before the demographic transition, with many offspring and another approach after the transition, with few. In some ways, there are similarities with the 'many' and 'few' strategies.

Before the transition, we have the 'many'-strategy. Humans evolved as a 'few' species, so we don't have vast numbers of offspring, but we need enough that our genes have a chance to survive. You can't have too many, you need to invest resources on your children, but you need at least replacement rate. From nature, we evolved to have relatively many children compared to what we have today in the developed world because it was necessary. The strategy is similar to that of turtles: lay a lot of eggs, so some offspring make it to the sea. In our case, it means having enough children that some survive the dangers of infections, unclean drinking water, predators in a dangerous environment, etc.

When we start the transition, we still have a high birth rate, we have the exponential growth, and we are following the 'many'-path to a population collapse. If we are in an environment where we can expand, and there are plenty of resources, this is excellent, but we usually aren't, so we have a 'many'-strategy where a 'few' approach would be better. If the strategy goes on long enough, there will be starvation and suffering, there will be competition over resources and likely wars, and by fire or by ice, the population growth will halt. The population might collapse into a smaller size, from which exponential growth resumes, or it can stay high, but with high mortality because of limited resources. Either way, it is a failure of the strategy. If there are limited resources, and the mortality rate is too low, the 'many'-strategy leads to suffering. The mortality rate will necessarily go up again to keep the population within the constraints of the available resources. An increase in child mortality won't necessarily be the way that the mortality rate goes up, but it will go up somehow. It has to. Things that can't go on forever don't. The situation must be resolved by one of two changes: increase the mortality or reduce the birth rate. The first choice leads to increased human suffering; the second choice is the last step in the demographic transition.

After the transition, we are in the 'few' strategy. Few offspring that we invest more resources on. Without population growth, the exhaustion of resources is no longer inevitable. This does not mean that we cannot run out of vital resources—we can deplete nonrenewable reserves—but that is a topic for the next chapter. For now, I will assume that a stable population size is safe, while exponential growth clearly is not viable.

Putting on our evolutionary glasses, we can examine what is happening here. The first stage of human existence goes back through all our species' history and numerous ancestral species. It is natural for us to have many offspring, even though we have to invest many resources in each, and this was necessary to propagate the species. When child mortality drops, we stick to the same strategy, although it is no longer viable. Nothing is surprising about this if we evolved to have many children. The surprising part is that we then reduce the birth rate. The change doesn't seem to be because a population reaches its limit, although some populations are pushing it. We *do* see societies with unrest and war when they suffer from overpopulation. We see nations break down when too large a fraction of the population consists of young males (of course) with little access to wealth or power. So we *do* see the mortality rate go up when we approach resource depletion, but we do *not* see the death rate increase high enough to match the new birth rate. Those populations keep growing. The growth doesn't stop before the birth rate goes down when we switch to the 'few' strategy, and it is because the birth rate goes down that the population growth stops.

Why do we switch strategy? Yes, from a human suffering perspective, we want a population size at equilibrium with the available resources, but evolution doesn't care about our suffering, it cares about the propagation of genes. If you send more copies of your genes into the future than I do, then your gene copies are likely to replace mine. If you have more offspring, then, all else being equal, you will outcompete me; if something in your genes makes your descendants have more offspring as well, then your genes will spread faster and faster into the future compared to mine.

Differences in the birth to death rate ratio will change the genetic composition of the species. If the population growth in one group exceeds the growth in another group, then this will change the frequency of alleles in the population. Alleles that are common in the first group will rise in frequency, while alleles that are common in the second will fall—simply because the fraction of humanity that is in the first group grows faster than the fraction that is in the second. We see this today; alleles from the developing world are increasing in frequency while those from the developed world are decreasing. This is a consequence of a slow spread of industrialization, wealth, modern medicine, and epidemiology. The demographic transition occurred first in the developed world, with exponential population growth there, but is now happening in the rest of the world. Some populations went through the middle phase of the transition early; others are just catching up. The shifts in allele frequencies could even out after the transition. Still, with exponential growth, one population that grows just a little longer than another will make up a larger fraction of humanity after they both stop growing. So the transition has changed and is changing the genetic composition of humankind.

From an evolutionary perspective, there is a benefit to the 'few' strategy if investing slightly more resources on fewer offspring leaves more descendants than the alternative of investing fewer resources on many offspring. It is better to be in stage three than stage one of the transition. In phase three, you can have two children, take care of them, and leave two descendants in the next generation. In stage one, you need to have more children to have two that survive. You still need to invest the same resources on the two survivors, but you also have to spend resources on those that die early, so overall, you are investing more resources in your offspring for the same outcome. In the middle stage, you can have many children, and they will survive. You might not be able to invest as many resources on them, but that matters little as long as they survive and pass on your genes to the next generations. From a quality-of-life perspective, it is better to be in stage three than in stage two when resources run out. But if you are in the exponential growth stage, then you want to have more offspring than the rest of the population; there is no selective advantage to be the one that switches to a lower birth rate first. You will just be outcompeted by those that do not.

The only benefit to having fewer children is if the extra resources you can spend on them give them a higher chance of reproducing than if you allocate fewer resources on more children. And that may not be the case—it could be the opposite—and yet we change reproductive strategy. The reason we change tactics could instead be that we have evolved traits, traits that were essential for us in the past, that now work against our evolutionary interests, and change our reproductive strategy.

We don't switch from a high to a low birth rate because of selection. The demographic transition is too fast; there is no possible way that the allele frequencies can change so rapidly that an entire population changes strategy in a century or so. So why do we switch? If I had to guess, then I would think the explanation for the change in strategy lies in at least two aspects of the psychology that we have evolved over many millions of years. One, we generally want prestige, and two, we want our children to do well.

When I say that we crave prestige, I do not mean that we all fight to become popes, prime ministers, or presidents. But we do want to stand out within our group. I am not claiming that you have that personality, most claim that they don't, and many are probably right. Still, the statistics say that most have this inclination—and there are good evolutionary arguments for why this is the case. We tend to be envious of friends and neighbors to a higher degree than we are of movie stars and billionaires. Not just about wealth or fame, but also simple things like attractiveness, stable relationships, professional or personal success. From a selection point of view, this makes sense.

We don't have to compare ourselves with people far outside our own group. We do not compete over mates with Bill Gates or Scarlett Johansson. If there is any envy there, it is mild compared to what we can feel for people in our own social group, according to several psychological studies. It is within our social circles we find mates. This is historically true, although things have changed somewhat with interactions in dense urban areas and with internet dating. However, we evolved in small tribes where we had to find our mates through our social circles, and our brains evolved to operate in those conditions.

Prestige makes us more attractive. At least there is strong evidence that for women, prestige is attractive in a man for a long-term relationship. For a short-term relationship, women usually prefer a dominant male, but it is long-term relationships that matter for this discussion. What men find attractive in females is mostly physical attractiveness. We are shallow, but to our defense, we evolved that way. Humans want to appear wealthy, powerful, and beautiful; we want to signal that we are living the good life with resources for exciting experiences; we want to show off on Facebook and Instagram; we want to look attractive, and this is how humans can show attractiveness. The term used in biology for showing off to attract mates is *courtship display*, and for humans, this is what we display. We want to accumulate prestige to attract mates, but probably not for the prestige itself. There is no evolutionary reason to want prestige if not to increase our fitness.

I want to stress that we are not talking about a pursuit of prestige to an unhealthy or sociopathic degree. Of course not. When you look around you, that is typically not what you see. It is not something we think about, either. We might feel a bout of envy, then it goes away, and we never think about it again. But we do want to show off attractiveness. You see it every time a friend puts up a picture of her dinner on Instagram or a picture of his holiday on Facebook. It is not only about sharing a pleasant experience with friends, but it is also signaling that they have a good life, and that makes them attractive. It is not something we are conscious about, but it is how evolution made us.

It doesn't turn off when we have a mate, or we already have offspring. Evolution wouldn't care if it turned off after reproductive age when it isn't useful any longer, but there is no damage done by keeping it going, and so no reason to evolve any off switch. There is every reason to keep it turned on when you have a mate and can still reproduce; a more attractive mate might come along. It doesn't matter if you can't have children, don't want to have children, or already have the children you wish to. Your brain is wired to want to show off certain things because, in the past, those things increased your chances when competing for mates.

It requires resources to project this perfect picture of yourself to the world. If you want to signal wealth, you usually need to have some wealth. You typically won't get away with bragging about going to a fancy restaurant if you never go out to dinner. You cannot impress your friends with your holiday pictures if you can't afford to go on holiday. We have a trade-off between gathering what we need for courtship displays and caring for our children. The purpose of the courtship display is mating, but our subconscious doesn't necessarily link the two. There is no reason to add rational motives to a psychological drive we need; it is enough that we have the drive. The purpose of sex is reproduction, but we have sex because it is pleasurable. Evolution needs us to reproduce, that is why we have our sex drive, but our sex drive is not limited to reproduction. We have recreational sex because it is sex we crave and not directly procreation.

When it comes to reproducing and showing courtship displays, we have two drives in play that are at odds. We want to look attractive, and we want to have children, and while the first only evolved because we need the second, our brain sees it as two separate pursuits and wants them both. We do not exclusively focus on gathering courtship displays and not procreating—if we evolved that way, we wouldn't be around any longer. But if we can have fewer children and still propagate our genes to future generations, then we free up resources for pursuing our other drive, gathering prestige.

We need to have children to propagate our genes, so our brain evolved to want children, but it isn't thinking about the benefits of having many children. That might be selectively advantageous, but it takes evolution time to catch up with a new environment. The brain wants to find a trade-off between children and prestige, and it could be satisfied with fewer children and then have more energy for accumulating courtship displays. There is no reason why our brain should focus on how many children are *born* as opposed to how many survive to adulthood. If the brain wants two grown children, say, then how many children must be born depends on child mortality.

If this were entirely subconscious, we wouldn't see an effect of lower child mortality. Our reptile brain doesn't plan ahead, and wouldn't see that its goal for having grown children is met before they are already grown. The strategy would remain the same, to have many children, and each generation, we would be surprised to see that many of them survived to adulthood. But we are not driven entirely by our subconscious; we are reasoning creatures and can observe the consequences of a change to our environment.

We can plan, based on our observations and reasoning. If we see that more children survive, then we can also observe that we can reduce the birth rate and still propagate our genes. We might be wired subconsciously to desire the latter, but we can change our behavior according to changed circumstances.

Say that, on average, there is a desire for about two adult children. If the mortality rate is high, then you need to have more children than that, but if it is low, you can get your two adult children from two babies. If this is where the brain focuses, then changing the child mortality rate changes the weight in the trade-off between children and prestige pursuits. With high child mortality, you might need five children to have two that make it to adulthood. With a low mortality rate, you only need two. If you still have five children, you get more adult children than you 'need' and the resources you have spent on them took away resources you could have spent on your courtship display. When the death rate goes down, the desire to have many children goes down, because it isn't a desire to have many babies but to have your children grow up. The brain doesn't switch strategy right away; it lives in a culture that, to some extent, determines how many children you have, but after a little lag, we get the change. When some people change strategy and have more prestige to display, others will see that display and emulate it. They still get the children they desire, but also the prestige they want, so that is preferable to prioritizing children over career. If prestige is our courtship display, it is part of our competition for mates, and if we see others with more powerful displays, then our instinct is to compete harder.

The second reason that I think is behind our reduced the

birth rate is our desire to provide for our children. Since we have had to nurse and protect our children through an exceptionally long childhood, our psychology has evolved to make us do this. If not, again, we wouldn't do it, and then our genes would be lost. The love we feel for our children was put in our brains because we need our children to survive and propagate our genes. The brain doesn't think this way, of course, it just produces the feeling of love, but that feeling is there for this purpose. The brain doesn't sit and think about reproduction and propagating genes; it is just wired with emotions that make us do it. And here's the thing: the drives for sex and love for our children live their own life. They aren't directly tied to the purpose they evolved to serve—reproduction and assisting our genes in surviving. Our emotions will drive us, regardless of how selectively beneficial or harmful they are, unless our psychology changes through some future evolution.

If we can see that we reduce the quality of life for our children, if we have too many, this will conflict with our parental love. We have the intellect to recognize that we no longer need as many children to ensure that some survive. We can also see that, within the limited means that we have, we can give our children a better life if we don't have too many. With the brain wiring we have, we can intellectually see that lowering our birth rate is an optimization of our evolved drive. As more people take this leap of logic, the culture will change, and society will change with it.

Our behavior is determined by how our brains evolved, and what drives and desires hide in our subconscious. Our intellect makes us capable of changing behavior when circumstances change, but drives are ancient, evolved over millions of years, and they do not change as rapidly. We evolved in a world with high child mortality rates. It appears that we switch breeding strategy when we enter a world with low child mortality, but this doesn't necessarily mean that it is an optimal strategy. There is still a selective benefit to having many children, as long as you do not lower their fitness substantially by having more, and I doubt that the difference between, for example, two or three makes much of a difference. If one group has 50% more children than another, say three instead of two, they will have left twice as many descendants after two generations and more than ten times as many after six generations. The growth can't continue, so the population will hit a maximum, after which the mortality rate must go up, but even here, there is a selective advantage to have many children. If we are assuming that people from the two populations die at the same rate, then we can look at the ratio of those that reproduce at a rate of three versus two. The former will be more than 98% of the population after ten generations if the two groups start with the same number of members. This is an *extremely* strong selection.

If we switch to a strategy with a low birth rate, it appears to be against our gene's interests. Other selected traits, our psychological drives, overrule us here; it is evolutionary baggage that hasn't adapted to our current situation. But the incredibly strong selection for more children will have an effect. When child mortality rates went down, we all shifted strategy to fewer children, but some still have more than others. If the populations in stage three of the demographic transition are stable, then if some people have fewer children than necessary to keep the population stable, others must necessarily have more. Bad for the alleles in the first group and great for the people in the second.

It will lead to a shift in the population's allelic composition; the genes for having more children will increase in frequency while those that prefer fewer children will decrease. Over generations, the 'many' genes will take over the population, and along the way, we take a step back in the demographic transition and enter exponential growth. All selection has to do, to get us breeding at a high rate again, is to nudge the trade-off between reproducing and our other drives, slightly towards having more children.

I think it is inevitable that future humanity will want more children, and the selection that results from even a minuscule higher reproduction rate is so strong, that I don't see how other drives can evolve in time to stop it. There isn't any benefit to reducing the number of offspring for an individual, and it is on individuals that evolution work, not society. It is the individual's ability to leave his or her genes in future generations that determine the selection, not some desirable situation for society as a whole. Even if society collapses, there is a benefit to outbreeding competitors. Fewer of your offspring will survive, but that doesn't matter as long as there are more than your competitors' offspring.

Is exponential growth and overpopulation, therefore, inevitable? Maybe not. We can imagine that people realize that having many children will lead to problems down the road, and their desire to protect their offspring from harm could kick in. But choosing for yourself to reduce the number of children you will have does not work. You will be outcompeted by those who have more children. To avoid breeding ourselves into a population collapse, we must fight against this selective drive, and it will have to be at the level of our entire society. To give your offspring a sustainable future hinges on the rest of society also limiting their number of offspring. Evolution works on individuals, but we are intelligent enough that we can override it and choose our future path collectively. Once a majority of society realizes that we must limit our birth rate, there will be growing social pressure to do so. There is not a strong stigma to having many children today, but if we approach overpopulation, there likely will be. And as humans, we don't want to be pariahs of society. Other people's disapproval could slow population growth. If nothing else works, there are more drastic measures. China's one-child policy worked to slow and stop population growth, and it can work again.

We will see a change in allele frequencies that increases our desire to have children because there is such a strong selective advantage to it. However, I also think that the resulting exponential population growth can be stopped in time before it is catastrophic. I return briefly to the issue of overpopulation in the next chapter, where I talk about our future interaction with the environment. The only thing that will stop us from exponential growth, if genes that prefer more children spread, is limited resources. At some point, we must limit population growth ourselves, or nature will do it for us. We do not want the second option. If larger populations are sustainable again further into the future, we will see an exponential growth resume. I return to this in the Going Forward chapter.

What happens if we put an upper limit on the number of children that people can have, whether through social pressure or law? A large part of selection is through variation in the number of offspring, and if we are forced to have the same number of children, could that halt our evolution?

Mutations would, of course, still enter the population, but allele frequencies can hardly rise or fall if we all send the same number of gene copies down the ages. The frequencies will change because we shuffle our genomes each time we produce eggs or sperm, so some alleles will be lost, and others will increase in frequency, but the changes will be slow if we all have the same number of children. If we limit the number of children we are allowed to have, then we slow down our evolution, but we do not eliminate selection entirely. We introduce an upper limit to the number of children each couple can have, but not a lower limit. Some will still have fewer children than others, and some will have none. Infertility or inability to attract a mate will always be there, and some will succumb to illness before they can reproduce. If there is less variation in the number of offspring, our evolution slows, but a one- or two-child policy is not going to stop us from evolving.

Aging

Our species evolved in small societies with high birth rates and high mortality rates. Most of the human population were children or teens, and few were old. Soon, the entire world will have transitioned to a world where a significant fraction of the population is old. At the global level, the estimate for 2050 is that 22% of the human population is older than 60. Already today, in seven countries, more than 20% of the population is older than 65; in order of percentages: Japan, Italy, Germany, Portugal, Finland, Bulgaria, and Greece. For Japan, the most aged population, more than one out of four, 27%, are older than 65, and Italy is not far behind with 23%. This is a dramatic change to our species, and it will affect our future evolution. When most of the population was young, genes that were helpful at a young age were strongly selected for. There will be genes that are beneficial at advanced ages, but in our past, few survived to benefit from them. The fraction of the population at an advanced age was small, and the selection wasn't particularly effective for the genes that matter there. Now, this has changed.

From a life expectancy below 30 in the Stone, Bronze, and Iron Age, we now have a life expectancy in the seventies and eighties in the developed world (and the developing world is not lagging far behind). Most of the increase in life expectancy is the decrease in child mortality, which strongly influences the mean life expectancy. Even in the past, it was common to live decades beyond the mean if you made it as far as the mean; the low life expectancy was a result of only a few children surviving beyond childhood. But even if we ignore the effect that child mortality has on the past versus current life expectancy, we are getting older. If you stratify by age, you see an increase in life expectancy for all age groups. We keep people alive well beyond reproductive age. Could some genes affect our bodies in our seventies, eighties, and nineties with alleles that were never before seen by selection? Can these be selected for if we live longer? The answer depends, of course, on how these genes affect reproductive success, either for ourselves or for our offspring.

Prolonged lifespan changes the composition of the human population away from primarily consisting of individuals at reproductive age and younger—those that will directly determine the genetic makeup of the future generations. In the future, the older fraction of the population will be the majority, and they will affect selection indirectly. If there are alleles that improve traits or quality of life after reproductive age, they do not directly affect fitness. Selection works on the alleles that affect the probability that you leave more offspring than others; if you have already produced the offspring you will provide, then evolution is done with you. Except, of course, if you have genes that you can use to improve the fitness of your offspring or your offspring's offspring. If we live long after reproducing, and stay healthy, we can assist children and grandchildren (and great-grandchildren if we extend life enough) in various ways. If the current strategy to have fewer children and then invest more resources on them works, then it is because the parents' investment in their children is worth it. If it works for the parents, it will also work for grandparents and older generations. Older generations can boost the fitness of younger generations if they stay around longer and are healthy longer. Genes that affect us in old age might increase the fitness of our children, grandchildren, or even great-grandchildren, indirectly.

Longevity is heritable, i.e., there are alleles that affect how long we will live. So there is a potential for selection. If investing resources on grandchildren and younger generations provides them with a selective advantage, then there is also a selection for living longer in order to do this. In all likelihood, there has always been selection for longevity, but when a small fraction of the population lived long lives, the selection was weak. Few people lived sufficiently long to be exposed to the selection, and even when they did, the randomness of diseases meant that their offspring might not live equally long. Your alleles might get a boost for one or two generations if you lived long enough, but then it could be several generations before those alleles helped another generation. Selection for longevity would have been exceedingly weak in the past.

Now, we all live longer, but people with certain genes live longer than average. These genes are exposed to selection, both because we reach older ages and can invest energy in further generations of offspring and because those offspring in all likelihood also live long enough for the longevity genes to see the selection. To provide our descendants with a selective advantage, we not only have to live longer, but we also need to be healthy longer, so here we should also see a selection. In our species' future, we should expect an extended lifespan with better health, and consequently higher quality of life, late in life. Medical technology will do much to effect this change, but selection will also play a part.

The generation time, the mean age we have when we reproduce, hasn't changed with the demographic transition; it is around 30 years in all three stages. The difference is that with many children, we have the first child earlier and the last child later, while with few children, we have children around the age of 30. The reason we have children around this age is mostly cultural. Those that have children very early or very late fall outside the norm, and falling outside the norm in a social species can affect our standing in society and thus our prestige—and we evolved to avoids this since it can decrease our selective fitness. Having children early doesn't adversely affect our fitness, though, even if it affects our social standing. We are reproducing already, after all. Having children early instead gives us a selective advantage. We can have more children, but even if we do not, there is a gain to having children early. If we breed above replacement rate, then the number of offspring grows with each generation. If the generation time is shorter for some people than others, the growth rate over time is faster. Think about the difference between 5% interest on a loan per year versus per six months. Losing status because we have children early doesn't decrease our fitness, as long as we have them, but we have evolved to avoid losing social status, and that will push us to avoid having children too early. Having children early is better than having them late; we are more fertile in the early twenties than we are in the late twenties and early thirties, where we tend to have children. And shorter generation times are better than longer generation times. Biologically, it is better to have children earlier than the current average, but our drive to avoid falling outside of the social norm makes us do something suboptimal.

It is not exclusively a social construct that we give birth at this age. We can reproduce relatively early compared to the mean, but fertility drops with age. Women's fertility peaks in the mid-twenties and decreases after that, drops dramatically in the forties, and reaches zero at menopause, typically in the late forties or early fifties. Male fertility does not fall to zero, but sperm quality decreases after around 25 years old, and the sperm cells accumulate mutations throughout a male's life. Hence, the risk of congenital disabilities increases with the father's age. Social pressures might give us a window for when we 'normally' have children, but the window is constrained by when it is biologically possible. Fertility quickly becomes an issue if social pressure increases the age at which we have children. I could imagine that the socially acceptable age to have children could increase, then that would conflict with our biological fertility, and that could, in turn, introduce a selective pressure.

Suppose that the social pressure not to have children when you are very young, is because we do not think that you are ready for it. Young parents have not made a career, and they have not accumulated the resources we feel are necessary to provide for children. This could be why, in the developed world, we consider teenage parents reckless, even though they are biologically ready for children. As the population grows older, the age at which we are deemed proper adults could go up. I don't think that the legal age of adulthood would go up; I don't see the voting age to increase or anything like that. But even if you can vote at age 18, say, the majority does not consider an 18-year-old quite an adult yet. That comes later, after you have finished your education, and after you have started making a career in your field of work. I realize that it is not all that get an education or have a career, but it is the trend of the majority, and it is the majority that determines when society considers you an adult. If the age at which you are considered an adult correlates with the age at which society finds it acceptable that you have children, then the pressure not to have children too early will remain in the future. If anything, it could go up. The decrease in fertility as we age could prevent an increase in the socially acceptable reproduction age, but considering that we already have children at an age where we are less fertile, it is conceivable that social pressure could dominate biological pressure.

Let's say that in the future, there is social pressure to have children even later than now. Infertility is already a problem for many couples today, and many now need in vitro fertilization to have children. If we increase the average

age of reproduction, then this issue escalates. Fertility later in life becomes essential for producing offspring, and a strong selective pressure appears. We cannot increase the reproductive age too quickly. Even if society changes, so we find it socially optimal to have children in the forties, too few will biologically be able to. But every time we increase the acceptable age for having children by a tiny amount, only a few will not be able to conceive at that age. If it is a tiny minority, it won't change the age that society finds appropriate, and they probably don't know that it will be an issue before it is too late—similar to how today, many couples do not discover that they have reduced fertility until the suboptimal age of late twenties/early thirties that is the socially acceptable norm for children now. Those genes that cannot conceive at the slightly older socially acceptable age won't make it to the next generation; the genes that can will make up the next generation. Generation by generation, the age at which fertility drops to unacceptable levels could increase. We would be able to have children in our forties and fifties. If the trend goes on, and our lives grow longer and longer, then maybe sixes, seventies, eighties, and so on, to wherever we can push our biology.

What about the pressure from the other direction? Fertility prevents us from having children late in life, but we can reproduce later than what we currently find the social norm. Society frowns on parents that have their children at advanced ages. At that age, the parents will have a career, probably made it far in that career since they have had years to do so, and they will have accumulated wealth to provide for the children. So why do we see it as reckless to have children in the forties or fifties? A common reason that people give is that we don't expect the parents to live long enough to support the children into adulthood. And even if they might live long enough for the children to reach legal adult age, they won't be around for the children as long as younger parents can. They will be old and feeble earlier in the child's life, so they might not be able to provide the support that society expects from parents. As the pressure on young parents is that they are 'too young' to support their children properly, the pressure on old parents is that they are 'too old' to do it. Both 'too young' and 'too old' are social constructs based on the expected lifetime of the population. A 30-year-old parent today can expect to live another 50 years. If a few generations from now, a 40-year-old parent can expect to live another 50 years, the argument for a parent to be 'too old' would shift. Now a 40-year-old will be around long enough to be a good parent. Generations later, if the life expectancy keeps increasing, 50 might be acceptable. The age could keep growing as long as medical technology, and our genes can keep up.

If the acceptable upper end of when you can have children increases, we can see another social effect that would increase selection for prolonged fertility. If the main reasons for having few children and having them in a short time frame are social pressures to gain prestige and show that we can care for our children, then opening the time window in which we reproduce can lead to more children. If you want to have many children, and let us assume this is still socially acceptable because we are not approaching overpopulation, then the primary pressure from society is that you are expected to provide for them. That puts a limit on how many you can have at the same time. But if you live long enough, and you are fertile long enough, then you can keep having children. When one child leaves home, you are ready to have the next.

The lower bound on when it is acceptable to have children could decrease. There is the selection for having children early, after all. But let us assume that social pressures are strong enough to prevent this. Then, if society gets more accepting of older mothers, it will only be biological constraints that prevent some from having more children than others. Society will accept that we have children at more advanced ages, so how many children we can have is only limited by how long we stay fertile. This will further increase the selection for fertility at older ages.

When we, as a species, grow older, we will see a selection for growing even older still, with a mean age more likely to be above fifty than below. If the selection persists, and I don't see why not, we can probably grow much older. There are biological reasons why we cannot extend our lifespan indefinitely; our cells are complex chemistry that will eventually break down. But selection combined with continuing improvements to medical technology could bring our life expectancy into the hundreds of years or beyond, depending on how much faith you have in future technology. We will be an older population, but hopefully without many of the drawbacks of old age, as we will also have selected for health and vitality in old age. We will likely have children later and over a longer lifespan, with several decades between siblings.

I can only imagine that the shift in lifespan and reproduction age will lead to many social and cultural changes. If we form our opinions and tastes in our youth and don't change them much later in life, as we do now, we could see a highly conservative society, perhaps with slower innovation. On the other hand, we will have a substantially more experienced workforce that could drive inventions even faster. We could grow more risk-averse if we expect to live for one or more centuries. I don't expect the rebellious youth to change in this regard, but the majority of society, which will be old, might. We might see more mixing of genes, indirectly through divorces. Those that divorce and remarry often want to have children in the new marriage but can be prevented from this by age. They might remarry late in life when fertility is too low, or they might not want new children if they have many from their respective previous marriages. If they can wait a decade or two and still have children, they likely will.

Culture changes incredibly fast compared to biological evolution and without as transparent a processes as evolution, so I hesitate to make too many guesses. I will bring up one opinion, however, because there is an interesting evolutionary angle to it. That is our next topic.

Sexual Selection

Sexual selection is a selection that works on mate preferences. You can think of it as a selection for attractiveness, only attractiveness in broad terms. If you have genes that make you more attractive to the opposite sex, then you are more likely to reproduce, and thus those genes have a selective advantage. The attractiveness can be physical beauty, but it doesn't have to be. Birds' songs or frogs' croaks are used to attract mates, and those with the most beautiful song or loudest croak attract the most mates, and so, sexual selection improves songs and croaks, even if there are no other selective reasons for those traits. They are courtship displays, they attract mates, but they do not assist the survival of the singer/croaker beyond that. Making noise makes you noticeable to predators, so from a survival perspective, birds shouldn't sing. They have to, though, because they won't breed otherwise.

When we choose a mate, our genes want us to select a mate that increases their chance of spreading through the population in the future. Our mate choice is wired in our brain, and as with many things, our brain makes us do for selective advantages, it does in a roundabout way. We don't explicitly consider the value of potential mates' genes; instead, we have a preference for what we find attractive. Attractiveness generally correlates with the quality of a mate's genes. We find health attractive, and our genes are better off if they live in a healthy body in the next generation. Wealth and prestige are attractive displays because those that have them can take care of our offspring—useful for at least one generation—and might have genes that helped accumulate their status, and we want to have those properties in our children.

Attractiveness doesn't have to correlate with fitness overall, though. It is obviously an adaptive trait, and thus fit in the evolutionary sense because we generally want to mate with people we find attractive. If our mate is appealing, then our children are more likely to be as well, and then they will find it easier to find mates themselves. In that sense, attractiveness equates fitness. But attractiveness is only a proxy for the quality of our genes, and we can evolve to be attractive without improving our genes in any other way; we can evolve false advertising, so to speak.

Selecting for attractiveness can come at the price of other desirable traits, traits that could improve your survival. If you are a bird and you want to avoid predators, then you want to stay silent and be camouflaged. Still, male birds usually sing and have bright plumage. If a bird does not call attention to himself, he will not find a mate, and then his genes are at an end. The long tail of a peacock is an extreme example. The male peacock is at substantial risk of predation because of his tail but is more likely to attract a mate. The individual bird would be better off without the tail, but he wouldn't mate, so birds with a short tail are less fit than those with a longer tail, in an evolutionary sense. Male peacocks have long tails because female peacocks prefer long tails.

If there is a characteristic that we find attractive, even if it is at odds with other desirable traits, we can see a strong selection for it. Who gets to mate, and who doesn't, obviously have a huge impact on future generations. Even if the attractive trait doesn't otherwise signal 'good genes.' Attractiveness is how our brain tries to recognize good genes, but it can be fooled. It can be fooled because it can be equally effective to fake the quality of your genes by displaying attractiveness with nothing to show for it, as it is to present good genes honestly. The brain's attractiveness mechanism can also be hijacked by completely random traits the brain happens to consider attractive. The same mechanism works whether attractiveness correlates with the quality of a mate's genes or not. Attractiveness could be entirely arbitrary, and we would still select for it.

There is a scenario called *Fisherian runaway*, where selection for attractiveness can lead to very rapid change in a species. If for whatever arbitrary reasons, the majority of females in a species prefer a particular trait, then the males with that trait will leave a disproportional number of offspring in the next generation; it is those with the trait that most of the females want to mate with. Now, suppose that the gene for the trait correlates with the gene for being attracted to the trait. Then, the male offspring of an attractive male will have the attractive quality, and at the same time, the female offspring will be attracted to the trait. In the next generation, there will be more males with the trait—because more males with the trait mated in the first generation. Also, a larger fraction of the female population will desire the trait—for the same reason: the males that had more offspring in the previous generation had daughters with the gene for desiring this attractive trait. Each generation has more of the attractiveness gene and more of the 'desire' gene, and very rapidly, the entire species will possess this particular trait and find the trait attractive. There will be an arms race for displaying more and more of the trait. We can see extremely rapid evolution this way, even if the trait is entirely arbitrary.

A trait that is attractive despite putting the carrier at a survival disadvantage doesn't have to be decoupled from displaying good genes, though. The trait can be arbitrary, but if it is generally a handicap to the attractive individual, it can signal the quality of his genes indirectly. If he can survive with such a noticeable handicap, then his other genes must be of high quality to compensate for it. We can evolve to find a handicap attractive in this way: arbitrarily, our brain fixed on some trait to find attractive, despite it being a handicap to those that had the trait. Of those with the trait, only the fittest survived to reproductive age. So those with the preference for the trait ended up mating with individuals with good genes. Their descendants continued having the preference for the trait, and they outcompeted the rest of the population because they generally had good genes. The handicap is a filter that guarantees good genes; thus, it evolved to be a display for fitness, and brains wire that as an attraction. The trait is attractive because our genes have evolved us into seeing it as a signal for the general quality of genes, *despite* the harmful properties of that trait.

With birds, it is the self-destructive singing and bright plumage, but I would argue that we humans also have self-destructive traits that signal fitness. Not in our physical appearance—generally physical attractiveness for humans signals health, so that is a direct signal of the quality of our genes. We humans have behavior that is self-destructive and could be a case of sexual selection.

In humans, we see behavior, typically in the teens and twenties, that is potentially detrimental to our health and survival. The behavior is more prominent in young men than young women, and this isn't unusual in the animal kingdom. The males are competing to attract females to a higher degree than vice versa—the same as for birds and frogs and many other animals. But the behavior only works because the females find it attractive, so the entire blame cannot be put on young men.

You might smoke because you think it looks cool, but you only feel that it looks cool, because you imagine that *the opposite sex* thinks that way, and there is a grain of truth to it. Smoking doesn't show that you have genes good enough to avoid cancer—you get it so late in life that it is irrelevant at the age when you are looking for a mate—but it shows that you are willing to take risks, and that can signal good genes. You might drive too fast, even though your risk of a debilitating or fatal accident increases, again because you want to show off, and it signals that your genes are good because you expect to survive taking risks. Extreme sport serves the same purpose. There are safety measures taken there that significantly reduce the risk, but you appear to take risks, and that makes you attractive. Taking risks is self-destructive, but it is seen as attractive to potential mates.

When young males get into fights, we see a touch of sexual selection. Their potential mate sees that they are ready to fight to protect them. But there is also another aspect to it. Fighting over access to mates is common in the animal kingdom, and we do it as well. It has served us well in the past; we are the product of genes from males that could fight off other males. Today, as a general rule, we don't have to fight with other males to mate, it is generally self-destructive behavior, but we still have the tendencies, because we evolved that way.

We usually grow out of this behavior, although some more than others. There are still those that are looking for a fight at age 50, but as a rule, you are more likely to get into a fight when you are 20 than when you are 50. As a society, we do not condone destructive behavior for those in the twenties, but we expect it to some degree. We are not as forgiving about destructive behavior from middle-aged people, because we expect them to have grown up and gotten smarter about it.

We have different behavior in different stages of our lives because selection needs us to do different things. When we are looking for a mate when we are young, we need to show off our genes. Once we have started a family, we need to protect our offspring. We can take risks to find a mate because we do not have much to lose. It is either finding a mate or not reproducing, so whatever it takes to look attractive is worth it. Once we have offspring, taking risks puts us at a selective disadvantage. If we die, we cannot protect and provide for our children. What we find attractive also reflects this. We are not, generally speaking, looking for cowardly genes in our youth, and we do not want reckless behavior in a spouse.

We start with the self-destructive behavior at a younger age than we reproduce today, but around the time that we are biologically able to. We usually stop the behavior before we settle down and start a family. Our reproductive behavior is slightly out of sync with the behavior we use to attract mates. At the age that we actually reproduce, rather than when we potentially could, we are attracted to different traits than those we evolved to display. If you keep behaving in your late twenties as you did in your teens, your fitness will be lower. Mates do not find it attractive. You have to grow up or leave the gene pool. If we start reproducing in our forties or fifties, teen-like behavior will put you at an even greater fitness disadvantage.

Since we, after the demographic transition, started reproducing in our late twenties and older, the genes that matter for attractiveness have shifted. Reproducing later will change them even more. Now, what matters, is our behavior from the late twenties to thirties; in future generations, it might be older. At this age, being calm, gentle, stable, prosperous, and so on, is what is attractive. Impulsive, aggressive, reckless are not attractive traits. So we should see a selection for the former and against the latter.

It doesn't necessarily mean that we will eliminate destructive behavior in youths. As long as they grow out of the behavior, they are not at a selective disadvantage when they reach reproductive age. But there is a correlation between behavior in youth and later years, so selecting for a change in behavior for thirty-year-olds could also change our teen behavior. It is possible that our different reproductive patterns can lead to a species that is less aggressive and less warlike. Wouldn't that be nice?

Immortality

To close the chapter, let us consider an extreme amongst the possible future demographics of our species, one that we cannot reach through biology but might reach through technology, a society where we are immortal. We cannot entirely defeat death; after all, accidents can happen. But we could imagine a future where fatal accidents are the primary cause of death, and where our life expectancy is more than a thousand years—and getting longer as we manage to make our environment safer and safer.

In such a society, we see the extreme consequences of an aging population, and by necessity, drastic measures to curb population growth. In this section, we consider the implications for reproduction.

We could imagine two scenarios for reproduction, one where fertility is prolonged with our longer lifespan and one where it isn't. The first scenario could simply be a continuation of the trend I described in the aging section. We grow older, and we reproduce later because societal pressures are the same, but nothing else changes. Today, we have children around age 30, and our life expectancy is around 70 to 80 years. Maybe we will reproduce at age 40 when we live to 100 years. It could then be around 400 years when we expect to live for a thousand years. There will be a wider window where we reproduce if we measure our lives in many centuries, but relative to our life expectancy, it could remain about a quarter into our life. If that is all there is to living longer, we will have slowed down evolution by a factor of ten if we reproduce at age 300 instead of 30. Selection works on changes in allele frequencies per generation, so the generation time is the fundamental unit of change. If we increase the generation time, we slow evolution by the same amount. It is not something we will notice, of course. Evolution works slow compared to our current lifetime, and it will also work slow relative to a life measured in centuries. We will not perceive a difference of a factor of ten.

If fertility still decreases rapidly with age, and we still need to reproduce in the first few decades of our life, the age difference between two generations could be limited to around 50 years. You will be about 100 years older than your grandchildren and 500 years older than your grand grand grand grand grand grand grand grand grandchildren. When you are 1000 years old, you can have 20 generations of descendants. Although they are your direct descendants, they will only share little genetic material with you—in each generation, they also get genes from other families, and that dilutes yours—but they will still be your family.

It might be hard to keep track of them, though, and family birthdays would be a logistic nightmare. If each couple is allowed two children under the laws in place to avoid population growth, then you can have two children, four grandchildren, eight grandchildren, and further doubling each generation. If you add up all the descendants you have in those 20 generations, you will end up with more than two million.

The number of descendants grows exponentially with a doubling at each generation, but the population size doesn't increase. Remember that other families also contribute to these descendants. Your two children have two parents, you and your spouse. Your grandchildren have four grandparents. Your grand grandchildren have eight grand grandparents. The number of ancestors grows with each generation at the same rate as your descendants in each generation. The population size is stable.

You won't meet much of your family more than a few times, if at all. You have two million descendants. Their extended family adds another two million (assuming that there are no overlaps in the pedigree, which there, of course, will be). There will never be enough time for all the birthdays, weddings, and baptisms you need to get to know all of them. You will probably only know a few generations of them personally.

Because you do not know them, the psychological imperative to help your offspring will be weak. The selective boost your descendants could get from a living ancestor, the effect we talked about two sections ago, won't involve you. You will boost the first handful of generations, perhaps, and then you will not be part of future generations' selection. You can directly affect selection for a century or two, but after that, your job is done.

It doesn't mean that a population where the majority are

centuries or millennia old does not affect human evolution. Such a society will be vastly different from what we know today, and norms will be skewed towards the older population's standards. If we live past a millennium on average, then most people will be a century or older. If we accumulate wealth and prestige throughout our lives, the wealthy and powerful will be in the older age brackets, and their goals and values will hold the most weight. The social environment we live in is very much part of our fitness landscape, and it will be strongly influenced by the norms and expectations of the aged population.

Could we get to such a society? How would we get to a stable population with early reproduction and long lives if available resources limit our population size? If we have a constant population size and then add longer lifespans, the population will grow. Each couple can have two children, so the number of people per generation doesn't increase, but there are now more generations alive. Each time we extend our lifespan, so we live to see one more generation, the population has grown by that generation's size. It isn't an exponential growth, so we are better off than if we increase the birth rate, but if our lifespan increases tenfold, then so does the population size.

When we live longer, that is, our mortality rate goes down, then we need to reduce our birth rate, unless there are sufficient resources for the population to grow. What happens if there isn't? And how do you deal with not knowing how long we can extend our life to reduce the birth rate in time? The population size a thousand years from now depends on how long the current generation will live, and they are just about to reproduce now. How many children can you allow them to have to avoid a catastrophic population collapse a millennium down the line? If we can have children later in life as we live longer, this is not a problem we need to worry about; every time life expectancy grows, we can move the acceptable reproductive window up to match it. We can stretch out the generation time, so we have the same number of generations around, just over a longer time. If we *have* to reproduce early in life, and we cannot predict the future population size because it depends on how long we get to live, then we need a more drastic mechanism to deal with it.

One possibility is explicitly to limit births to match deaths. Let people sign up for a reproduction permit and hand them out—first come, first served—as people die. Inhumane, perhaps, but from a utilitarian perspective, it is better to prevent some from procreating than to let all humanity suffer from overpopulation centuries later. It forces the birth rate to match the death rate; you get one birth per death, no more and no less.

If the population is at equilibrium, the birth and death rates will roughly match, and using a queue for birth permits will not be vastly different from limiting the maximum number of children per couple to two. But in a phase where life expectancy grows, not everyone can have children. Who gets to propagate their genes will depend on who is admitted to queue up and for how long they can stay in the queue.

If everyone can apply for a reproduction permit, and if people can stay in the queue as long as they are fertile, then you increase your chances if you sign up early and remain fertile longer. We already discussed selection for extended fertility, and I expect that before we get to extreme life extension, we will already have seen it. Waiting in line will only strengthen this selection. There might be the requirement that you have to sign up as a couple, and you could be thrown out of the queue if you break up. In that case, there will be a selection for entering relationships early, so you can sign up early, and staying together long enough to have children. We could end up selecting for stable relationships.

If you can sign up for the birth queue as an individual, there is an asymmetry between men and women. Both can hope that they are in a relationship when their number comes up, but if not, a woman can be artificially inseminated. A man can, at best, be permitted to donate sperm. I don't imagine that donor sperm is selected in chronological order from the men that sign up for the queue. And in any case, I don't see how you would force shared custody of a child between a mother and a sperm donor. More likely, donor sperm is selected the same way as it is today, from a profile of the donor. If so, and if women are allowed to sign up for artificial insemination, then whatever is considered attractive in those donor profiles will be selected for. One sperm donor can potentially have numerous offspring, even under the restrictive birth regime, and we can see an exceedingly strong selection for some traits.

There is no guarantee that the birth queue will be fair, however. We might not let everyone sign up. It is an ideal setup for eugenics, the improvement of the human gene pool based on desirable properties. We breed plants and animals for traits that we desire, and nothing except for morals prevents us from doing the same with humans. If we already restrict people's right to have children, it is a small step to select who can reproduce from their genetics or phenotypes. If we only allow a minority of people to have children, we could take the opportunity to get rid of some genetic diseases—screen for them when people sign up and don't allow those that carry them into the queue. We could go further and select for other traits. Put a lower limit on the IQ, and if you fall below it, you don't get to join the line. Whatever property we desire, once we control people's reproduction, we can select for it.

Eugenics sounds horrible to most, especially since it is strongly associated with Nazism, but it doesn't mean that a future society will refrain from it. When someone selects a sperm donor, they are choosing a phenotype based on their preferences. It is a form of selective breeding. We accept selective breeding, even for humans, in some circumstances. We do it with donor sperm, and I will come back to other cases later in the book. The ethical issue with eugenics is not, for most people, actively selecting traits we want in our children when we get to choose. It is preventing someone who wants children, from having them, based on characteristics and genes they possess. If we are already selecting who can breed to avoid explosive population growth, then maybe the small additional step to selective breeding is not that large. A side effect of living longer might be that we expose our species to intense artificial selection.

I am not suggesting that this is desirable, and I do not think that it will happen—I think we will artificially change our genes in different ways, and I will return to that later. But I do see it as something that might happen if a future society needs to balance the birth rate with a decreasing mortality rate. If we only allow some genes to be copied to each generation, we might as well choose the 'good ones.'

If a government enforces restrictions on how many children we can have, there will be people breaking the law. If genetic factors can make you more likely to break the law in this particular case, then we have another potential for selection. Genes for being outlaws, anyone? If there's the death penalty for those that do not respect the reproduction policy, that selective advantage can quickly turn. A bit draconian, yes, but we are looking at a dystopian scenario of what could happen if we need to avoid overpopulation by force.

But let's say we have a stable population where we have reduced the risk of accidents to a level such that people live for tens or hundreds of thousands of years. If the only cause of death is fatal accidents, there will be a considerable variation in lifespans. If the mean lifespan is a hundred thousand years, then the oldest members of societies might be a quarter to half a million years old. There could be ten thousand generations between the oldest and the youngest members of the species. Our species is substantially less than half a million years old, so we are considering a population where the oldest members could be viewed as a separate species from the youngest generation.

With strict control over reproduction, we will see a slow evolution, but over hundreds of thousands of years, change is inevitable. Allele frequencies change slowly in a population where people stay around for millennia, but the genetic composition of younger generations drift away from their elders. What is interesting to consider is whether different generations, living side by side, can have different evolved traits.

Our ancestors from half a million years ago, were a different species. They are lost to us through time, but in our distant future, ancient ancestors could still live among us. There would be slight physiological changes. Not many, we don't evolve that fast, but there would be some. We could have diverging drives and desires from modifications to our psychology. In this sense, we could see a society with a progression of different species living side by side. I wonder what senior citizens will think of 'the youth today' a million years from now.

CHAPTER 5

In Our Natural Habitat

All living organisms affect their environment to varying degrees, but we humans, through our intelligence and numbers, do so to an exceptional extent, and unfortunately, often with negative ramifications. We burn fossil fuels that pollute the environment we live in and change our atmosphere leading to climate changes. We destroy the habitats of other species, leading to extinction and a reduction of the worlds' diversity. We also impact our environment by building our own habitats, changing the environment for our needs, for example, by building houses and cities.

We are not adapted to the environment that now surrounds us. How we modify our surroundings didn't slowly change as we evolved as a species. Neither we nor our fellow life had time to evolve along with the changes. We affect our environment, not through evolved behavior, but through our inventions and ingenuity. Beavers developed to build dams, they are adapted to the environment they create when they make a dam, and other animals that live in their habitats have no problem with dams. Eucalyptus trees produce oil in their leaves that increase the risk and intensity of forest fires, and while many plants and animals die in those fires, they do not cause ecological collapse. Plants and animals living along with eucalyptus have adapted to frequent forest fires. We didn't evolve to build cities and cars, and we are not adapted to city life. We didn't evolve to do strip mining or burn forests to create farmland, and the species impacted by this behavior certainly didn't evolve mechanisms to cope with it.

The success we have had with adapting the environment to our needs enabled us to colonize the entire planet and to grow to numbers otherwise unheard of for large mammals. Well, almost unheard of; some other large mammalian species have population sizes up to a billion, but we can take credit for that as well—they are all domesticated animals and owe their population numbers to us. We have multiplied and expanded at an extreme rate and changed everything in our path. We have done this, not as a consequence of evolved drives that make us do these things intuitively, but as a side effect of our critical evolutionary invention, our intelligence. We can react to new conditions, we can identify and solve problems, and we can change our environment to fit our needs, rather than wait for evolution to adapt us to our environment. But the speed at which we have changed the world has not given evolution the slightest chance to adapt us to the world we have created.

We have utterly changed our environment and, thus, our fitness landscape in a blink of an eye, with no chance for evolution to catch up. We have changed the world to meet our needs, so we are not maladapted to it. Whether we change to match the world or the world changes to match us, it doesn't matter much. We still fit well together. However, what we have done, we did for immediate and obvious gains, and in many cases, we did not know what side effects and consequences, the changes could have both to ourselves, and life on Earth in general. In some key aspects, we are a good match for the world we have created, but in others, we are not. When we changed the world, we did it so that it met some of our needs, but without considering ill consequences that the changes also had for us. We have many attributes that are not adapted yet, and we have just started our journey through the new fitness landscape.

We need to adapt to urban life in a world with high population density. We need to learn to live within our means when our sheer numbers strain the world's resources. We must learn to interact in new ways in a global society with instant communication and high-speed travel. As we change the environment around us, we must adapt too.

Urbanization

We evolved to live in small bands of 100–200 mostly related hunters and gatherers, but urbanization is changing this. Urbanization is not a new phenomenon—people have lived in cities for thousands of years—but the size of cities and the number of people moving into them is increasing at a hitherto unseen pace. Half of the world's population now lives in urban areas, and the process of urbanization is accelerating. Fewer than a billion humans lived in urban areas in 1950, while more than four billion do so today. A factor in this is that the worldwide population growth is mainly in urban areas, but there is also a massive movement of people from rural to urban areas. In the future, most of humanity will live in megacities, cities that are the result of smaller cities merging as they expand into each other, cities with tens or even hundreds of millions of inhabitants. It doesn't necessarily mean that we will eventually cover the entire Earth in cities, although we could approach such a situation if our numbers keep growing. If we keep the population growth low, the cities' growth will be slow. If the population growth stops, then soon the growth of cities will as well, even if people continue migrating to cities for a while after the world population stabilizes. But before we reach the point where cities stop expanding and merging, the vast majority of humanity will be urban.

We are not forced to live in cities because there are no rural areas to live in; we are drawn there because there are benefits to urban life. It is where there is the most diverse range of goods and foods available in shops; the labor market is better than in rural areas; there is better access to education, and should you need a hospital, it is much closer. At the same time, we are pushed away from rural life as there are continuously fewer jobs available there. Before the Industrial Revolution, a significant fraction of the population was needed to farm, to feed the urbanites that didn't produce their own food. Then farming technology started to improve drastically during the Industrial Revolution and is advancing to this day. With better farming technology, relatively fewer hands are needed for agriculture, even when feeding a substantially larger urban population. Fewer still are required in the future.

There are disadvantages to living in cities as well, though.

Pollution, for example, and the psychological alienation from living with millions of strangers instead of a small group of family and friends. Still, the flow of people towards cities, rather than in the other direction, indicates that the advantages must be perceived to outweigh the disadvantages. The fraction of humanity that lives in urban areas will rise in the future. The population density will increase, and we will have to interact with more people than we have evolved to do.

We evolved to be a social and cooperative species. Cooperation improved our chance of survival, and groups better at collaborating would have an advantage when competing with other groups, thus giving their genes a selective advantage. Consequently, we are all well adapted to interacting with other humans, just not with as many as we interact with today.

In the past, most of our interactions would be within a small familiar group of people. Urbanization vastly increases the number of people we interact with daily. Of course, even if we live in a city with millions of inhabitants, and have online access to the billions composing all humanity, we will not interact with all of them. But we will still regularly interact with thousands using a brain designed to cope with hundreds. And it does seem like the brain is made for a large number of social contacts. The number of people we can have close and stable relationships with, called the Dunbar number after the anthropologist Robin Dunbar, is 100–200 with an average of 150. Not coincidentally the size of groups we evolved to live in. We have interactions with more, we can have more acquaintances, but this seems to be the number of people we have regular social contact with. It is not a static group, people enter and leave your group of contacts, but the group doesn't grow; you don't accumulate close contacts. Our brain seems to be able to keep track of this many people, but not more.

Considering the known benefits of having large networks, I would expect that being able to have closer relationships with more people would give you a selective advantage. I would expect us to evolve towards having substantially more close contacts and keeping in touch with them, so we don't drift apart, even when there are 500 or more friends rather than a measly 100–200. Having a larger group of friends means that you have a more extensive social network that can help you when you need them to. It gives you a selective advantage. We cannot keep track of everyone we are introduced to, in all social gatherings, but maybe we could get better. Imagine a politician or a businessman that is slightly better at remembering details about those he meets. Meeting them again, he has a closer bond to them. This will give him or her more support or better deals, which lead to growth in wealth and power. Power, prestige, and wealth attract mates, so this would be an adaptive trait. In the social groups that we evolved in, there was little use in a brain that could handle larger networks because there just weren't that many people you could interact with. Now there are, and it is perceivable that we will evolve to have many more friends.

Paradoxically, living in a city with people all around us can make us feel more alone than if we live surrounded by a few. In a small community, we interact with the same people repeatedly, and over time we get to know them and find friends among them. When you live in a small village, you know your neighbors; when you live in a big city, you don't even know the names of everyone in your block of flats. Living in cities makes us feel alienated and lonely. Those who find it hard to make friends under normal circumstances find it harder still when most interactions are with strangers. It takes more effort to actively try to find friends than simply to grow up with people. The psychological stress of being surrounded by people and yet haunted by loneliness can lead to depression and suicide.

Clearly, depression and, especially, suicide are not adaptive traits. Selection will work hard at getting rid of them, and either adapt us such that the feeling of urban loneliness does not harm our psychology, or it will change us such that we get better at socializing and finding friends.

We did not only evolve to collaborate; we also evolved to compete within our groups. Selection means that all living organisms have competition within their species. It is within our species that we compete for having more offspring than others. Since we are a social species, some of that competition comes in the form of alliances, cheating, manipulating, and lying. Sad but true. As we evolved and our intelligence grew, we were in an arms race against other members of our species in perfecting our social skills. In such a competition, if you are better at guessing other people's intents and schemes, you have an advantage. You can identify those who will reciprocate collaboration and those who are likely to take advantage of you, whom you can make alliances with, to protect yourself and defeat others, whom you expect to cheat, and whom you might get away with cheating. Everyone has to get better at guessing other people's intentions, or they will be outcompeted, so every allele that gives a slight advantage is selected for, and the selective pressure is immense.

It would be nice if urbanization only leads to more friendships, but access to many strangers have a dark side as well. The negative consequences of lying and cheating are vastly reduced. If you live your entire life with a small group of people, and you through your actions have made yourself a pariah to this society, you are in for a rough ride. If you have done it before reproduction, your actions made you evolutionary unfit. You would have to leave the group and join a new one to recover from your deeds—and if you were a repeat offender, you might never be able to settle down with a stable group to start a family. Deceiving strangers, people you will never meet again, might not incur any of the drawbacks of cheating. Being good at lying and cheating might even work to your advantage.

There are benefits to outwitting someone. You can cheat someone out of food or property and be better off yourself. You can manipulate others to obtain power and wealth. These are deceptive traits that we developed because they give us a selective advantage. Since it is a selective disadvantage to be manipulated and cheated, we also developed defenses against this.

Despite what many believe, we are not good at detecting lies. We cannot readily determine if someone is trying to deceive us. Police work would be so much simpler if we could, but even with technology like polygraphs, determining if someone is lying is unreliable. When we can be confident that someone has been lying to you is when we can detect inconsistencies in a lie, or when a falsehood does not match observable fact. We cannot identify dishonesty the minute we are exposed to it, but we can often identify lies when we can examine them over an extended period.

Our prime defense against manipulation is not in catching it in the act but retaliation when we have suffered from it. When we feel wronged, we want revenge. In a civilized society, we say we want justice, but this is usually a euphemism for revenge. We want to punish those that have wronged us, and we want to set an example to prevent others from trying it in the future. It doesn't only manifest with crime and punishment. We have a strong feeling of fairness, and we will go to great lengths to obtain it, even if it is detrimental to ourselves.

A classic experiment that demonstrated this is the *ultimatum game*. One player is given a certain amount of money, and it is his job to split it into two parts, one part for himself and the other for another player. The other player can then accept his cut or reject it. If he takes his cut, both players get their respective pile of money; if he rejects it, none of the players get any money. If you are the second player, you always get money if you accept the amount you are offered, assuming it is not zero, and you never get anything if you reject it. If you want any money, you should always accept. But people do not always accept what they are offered. If they think they are treated unfairly, they reject the offer. Apparently, it is a big part of our psychology to punish unfairness. We would rather forgo money to punish an unfair first player than to gain something ourselves. Our sense of fairness is so strong that we often prefer to be fair to others, even when there is no benefit to that. If you remove the second player's choice in the ultimatum game (it is then called the *dictators game*), the first player can safely take all the money because the second player cannot retaliate. But most players still divide the money to give some to the second player. There is no risk of retaliation, and it will always be economically optimal to take all the money yourself. Still, our desire for fairness overrules us. Certainly not always, people do try to optimize their earnings at the expense of others, but it is far less than what we would do if we were only interested in our own gains. A sense of fairness is not only a property of the human psyche, but many social animals also share it, so this is an old evolved trait. Fairness is apparently a powerful driver in social animals.

Someone who mistreats others will not be treated fairly in return; the social group will not trust someone who manipulates and cheats. This keeps the cheater genes and the selfish genes at bay. Yes, we still have dishonesty, lies, manipulation, criminal behavior, and such, but not nearly to the degree we would have if there were no adverse consequences. We have those traits because we evolved in groups where there were benefits to them, but we also developed defense mechanisms to protect ourselves at the same time. Our societies are predominately based on our defenses. Even though the acceptance of unfairness and corruption varies greatly, no society considers lying and cheating virtues.

Retaliation only works if the cheater and the cheated are in the same community. We cannot detect the lie in time to stop someone from cheating us, and if he is far away when we discover that we were fooled, then he gets away with it. Living in larger communities, where one can hide in the crowd, makes it easier to get away with cheating. Living with many strangers makes it possible to cheat new groups of people regularly because they will not be warned of your behavior. We have police to catch and imprison criminals, and they often have resources to track down con artists that the community he cheated could not find themselves. But they are nowhere as efficient at stopping a swindler as the entire community when the lier was part of it.

Even if there is effective policing, and if all downright illegal activity could be stopped, you can have an advantage if you are an effective manipulator. Not all cheating and not all unfairness is unlawful. You could start a company with someone who decides to pay himself twice as much in salary as you get, and there is nothing illegal in that. It is just unfair. A mechanic might tell you that he can fix your car cheaply, but not tell you that all the other local mechanics would do it cheaper. It is not a mechanic you would want to use again, but it is not against the law. Politicians will happily promise you everything you could ever want, knowing full well that those promises will never be met.

If we cannot retaliate with social pressure when we are manipulated and lied to, then we don't have the defense mechanism we need to avoid being cheated. Since there is an obvious benefit to exploiting other people's labor by tricking them out of their resources, and since the social retaliation that would nullify this benefit is gone, there is fitness to lying. It is not a new fitness, it has always been there, but without anything working against it, the selection is stronger.

But this was always an arms race between deception and manipulation on one side, and defenses against being cheated on the other. If a more substantial part of society starts cheating, then selection for defense mechanisms kicks in with equal strength. If we cannot retaliate when we are wronged, we must learn to detect liars early to avoid being scammed in the first place. Maybe we can get better at detecting untruth through body language and microexpressions. If not that, will we get better at recognizing when something is too good to be true? Or clashes with everything else, you know? Or isn't corroborated by other information you have available? It is not hard to recognize spam, conspiracy theories, internet trolls, alternative facts, and fake stories if you are a skeptic when you receive news. Most of us do not find it hard to recognize that a Nigerian prince who wants to send us loads of money is an unlikely thing to happen. We know that it is spam. But there are surprisingly many that fall for it every day—otherwise, we wouldn't see attempts of this scam anymore. Most of us, and I hope that includes you as well, are not fooled by 5G or Flat Earth conspiracies. And if you are like me, you probably check several news sources to see if an incredible story that pops up is reported elsewhere, and thus less likely to be false. How well our bullshit filter works varies, but we all have a filter to some degree. Still, no matter how careful we are, we can be tricked every time we let down our guards. If more people try to cheat us in the future, then we need to boost our filter of lies to avoid being fooled.

It isn't just our psychology that will be affected by urbanization, and it is not only through selection that our genetic composition will change. Humanity will interact more, our genes will blend more, and ethnic differences will diminish.

Cities tend to be more cosmopolitan, and urbanization tends to bring together people of different ethnic and cultural backgrounds. We usually mate within our ethnic, cultural, or religious groups, but there is always gene flow between groups. As a species, we have even interbred with other species of humans in the paleolithic, at least the Neanderthals and Denisovans, but maybe others as well. Group identity is not an impenetrable barrier for us; it might slow gene exchange, but it certainly doesn't stop it. The more groups we put in close contact, the more interbreeding we will see, and ethnic distinctiveness will decrease. The genes that make us different from other groups will not disappear, not for a very long time, but our ethnic distinctiveness is not caused by differences in genes, but by different allelic compositions. Our genes work together to create our phenotypes, and it is the prevalence of various alleles in different ethnic groups that make us appear different. A child of two ethnically different parents will have 50/50 genes from the two ethnic groups, but there are no 'half alleles' in there; there are no mixed genes. The same alleles are simply combined differently. We will not lose our genetic diversity when we start mixing more, but phenotypic diversity will be reduced. We will look more alike, and one could hope that this would lessen jingoism and racism. I doubt that in-group/out-group thinking will ever disappear, but we will have one less thing to base it on.

Mobility

Urbanization will mix our genes, and a trend with a similar effect is increased mobility. It has never been easier to travel, and with wealth increasing globally, everyone will be able to go anywhere in the near future. Globalization and international corporations increase our mobility as well. Companies are fighting to headhunt the best employees, and the best people are not always living in the same country as the company that wants to hire them. To be competitive, companies need to hire internationally. Today, all countries have laws that restrict immigration, but most also have exceptions for highly specialized and highly skilled labor. Nations also want to be competitive and need to allow migration for the same reasons that companies must hire internationally. The backside of this coin is brain-drain. Wealthy countries import from the best-educated population of developing nations; poor countries pay for their education but get nothing in return. As wealth broadly increases everywhere, this will become less of a problem, and the movement of people will not be one-directional.

Globalization and high mobility inevitably lead to genetic mixing. We find partners amongst the people we interact with, and if that is an international group, then we are likely to find a spouse from a different part of the world. I work in academia, where universities compete to get the best people internationally. I worked abroad, during my Ph.D. studies and my postdoc years, on three different continents. Several of my current colleagues are foreign-born, and more than half of my students are foreign as well. I have numerous friends married to someone born in a different country, and in many cases, different continents. If it weren't for the current CORVID-19 pandemic, I would be married to a foreigner myself, although not an academic. We will marry as soon as we can travel again.

The international community in academia leads to a high degree of genetic mixing. It is not, by far, the majority of academics that marry someone of a different nationality—the majority finds spouses outside academia in any case—but it is prevalent enough that no one thinks twice about it.

I don't have much experience with international companies and the mating habits there, but I suspect it is similar to academia. They hire their people from top universities, they go for the best qualified regardless of nationality, and they mix their staff from all nationalities. I have friends working in high-tech companies that are married to other nationalities, and I imagine that the situation there is very similar to academia.

It is not just headhunted, highly skilled people who will migrate and settle in different parts of the world. Aging populations are likely to need immigration from younger populations and to import people from elsewhere. The birthrate is dropping across the globe as everyone gets wealthier, and as better medicine becomes available to everyone, the fraction of older people will grow, as we have discussed earlier. Import of young people to compensate for an aging population will not stretch far into the future; we have to solve the 'problem' of an aging population through technology. Until then, however, we will see substantial migration—obviously more so than what we see from a few high-skilled workers. These migrants are more likely to settle in their new country than those who will move around from company to company chasing the best position and highest salary.

If climate changes get as bad as it could look like then we will see massive migration of refugees on top of this, so a vast number of people could move from one location and culture to another soon. Whenever migrants settle in a new country, they are likely to mix with the original populations—both culturally and genetically; not immediately, perhaps, but over time.

The higher mobility will increase the mixing of ethnic groups on top of the mixing consequence of urbanization. This is one of the reasons that I allow myself to consider the entire human species as a single population in this book. In the past, the human species was highly structured, into different groups with different allelic compositions. In the future, with higher and higher mobility, this structure will erode. There will still be variation between people, of course, but as ethnic groups mix, the population differences will eventually disappear.

Attracting Mates

Changes to dating patterns also speed up genetic mixing. A century ago, we mostly had to find mates via our social circles. You found a spouse through your friends or family, or your family did it for you, and that meant that marriages were predominantly within social groups. This is not only a rural phenomenon but also the pattern in urban societies. But now we have online dating, and online dating is replacing more traditional approaches to finding partners. A study from 2019 [1], based on a 2017 survey, shows that in the United States, finding dates online surpassed getting introduced through friends in 2013. By 2017, almost 40% of all couples met through some form of online dating. Finding a mate online is more popular than any other approach, and the fraction of couples that meets this way is growing steeply. Algorithms, rather than families, are the new matchmakers.

---

[1] Disintermediating your friends: How online dating in the United States displaces other ways of meeting, PNAS September 3, 2019 116 (36) 17753-17758

If you do not need friends and family as intermediaries in finding a partner, you can cast a wider net. You are more likely to get in contact with someone outside your cultural, religious, or ethnic group. Our feeling of group identity still limits how freely we find partners, but technology lets us find partners with different backgrounds than our own, to a larger degree than we could before. In most cities, different ethnic groups live side by side, but they do not mix much, because they mostly keep to themselves. When the dating scene moves online, the algorithm that now pairs us up doesn't necessarily look at your ethnic group, so it is happy to try to match people with different ethnicity. The computer opens a communication channel between otherwise isolated groups—and love follows. This is speeding up the merging of ethnic groups. Mixed marriages are on the rise, and we are also in this way, slowly eroding our differences to become a single population.

The computer also enables you to meet someone on the far side of the globe. Except for time zones, there isn't much difference between chatting with someone next door or someone on the other side of the Earth. Long-distance relationships are more complicated than relationships where you can regularly meet in person, but they still often lead to marriages and relocation. As international travel gets cheaper and more convenient, distance is less of an issue. Thus we can see an increased mixing of groups, reinforcing the effects of urbanization and (local) online dating.

On top of mixing populations, online dating can also affect mate selection. We choose mates based on a variety of clues, such as physical attributes, personality, and courtship displays. These clues evolved in a world where we could evaluate them over an extended courtship. When dating moves online, many of those clues are gone, and others are distorted.

When you scroll through a list of profiles, you cannot evaluate someone's personality. Anyone could have helped them write their profile text. They might have managed to write an engaging profile by spending an excessive amount of time on it, but never be able to communicate this well in real-time. They might sound witty in their profile because they borrowed jokes from elsewhere while being bores in real life. The closest to the truth you can get is their profile pictures, and that is assuming that they are of themselves. Some use false photos, of course, but that is easily detected the first time you are on Skype or FaceTime, so let's ignore that. But even authentic photos can be slightly dishonest. They can be Photoshopped or years out of date. Even if there is no foul play with the pictures, they will be chosen to present the potential mate in the best possible light, and that can also give you a distorted view of the person. Some are less photogenic than others, and that puts them at a disadvantage on the online dating market. Some might have great personalities, but if they do not look good on photos, they will be ignored.

It is not that you can indefinitely distort how you are, and whether you are an attractive mate, just because you are dating online. Eventually, you will meet, and once that happens, we have all the evaluation clues we evolved with. But there is a screening taking place before you meet someone. Only those that have attractive profiles get to show off their true selves on a date (for good or for worse). It is a filter that was never there before, but now can be essential to pass for reproductive success. If you do not pass the initial screening, you don't get to date. If you manage to show off an appealing profile, you have a shot.

Writing an engaging dating profile is a skill that you can learn, but as with all skills, some will be better at it than others. And it is a different skill from making a good impression on a first date, so online dating shifts the skills set you need to attract mates. If couples write together before they call each other, then writing skills are more critical than earlier, and your skills at making small talk might be less relevant. If the main criterium for picking people to contact is profile pictures, and as a species, we already put substantially more weight on visual impressions than any other sense, then looks will carry even more weight than it already does.

Which traits will be necessary for mate selection in the future, and what it means for selection, is hard to guess. If online dating is here to stay, looks and communication skills will matter even more in the future than they do today. There has always been selection for both, and for physical appearance, I don't think there will be any major changes. For communication skills, however, there could be a change in selective pressures, because we change how we communicate. Writing is different from talking, and for a shy person with excellent writing skills, for example, dating will be easier. When you are hiding behind your screen and communicating through writing, you don't display the same signs of nervousness, and you can correct your stammering before you hit Enter. An extrovert, with the guts to approach a potential mate and the self-confidence to strike up a conversation, has a selective advantage over an introvert when dating in a physical setting. That advantage could disappear in an online environment. Shyness doesn't have to be a handicap in the dating world of tomorrow, and showiness doesn't have to be an advantage.

If we filter potential mates based on online pictures, we are turning into a shallower society. Paradoxically, though, the dating scene where we would do this will probably also make us value deeper conversations more if we have a period of online communication before we meet physically. We might have two separate filtering layers when looking for a date, one based on appearance, then another one based on meaningful conversations. Only after these two filters do we continue with physical dates and all the evaluation machinery we evolved with.

Whatever the effect that a changing dating scene has, it will be profound. Mate choice is a core mechanism in evolution, and we are making dramatic changes to it. It is unavoidable that strong selective pressures follow.

Pollution

The energy and material we use and produce are polluting in minor or major ways. Some pollution has been with us for a long time—we have used fire since before we were Homo sapiens—and some pollution is new to us—we didn't have plastic until the twentieth century. What is pollution doing to us, and will it have evolutionary consequences?

Consider air pollution. It is estimated to lead to congenital disabilities and to kill about seven million people every year (causing heart diseases, respiratory disease, cancer, and more). According to WHO, 1.7 million of those deaths are children under the age of five. As a species, we are exposed to two main kinds of air pollution. One kind in the impoverished world, where many rely on wood stoves and oil lamps, and people are exposed to smoke—another type in the affluent world where industry and traffic pollute our cities with smog.

We have lived with air pollution for a very long time. We may not think of prehistory as a period with pollution, but if you are cooking over a fire, you are exposed to what is basically second-hand smoking. We haven't been smoking for long in evolutionary terms, and the fraction of the population that smokes is decreasing with the growing social stigma there is to the habit, so I will not discuss it further here. With cooking over a fire, you are not exposed to the same chemicals that there are in tobacco, but it is still unhealthy to breathe in smoke. We do not know precisely how long we have used fire, but it is more than a million years. So maybe we are, at least partially, adapted to it. Not using fire would definitely reduce fitness; it keeps predators away and is essential for keeping warm in temperate and arctic climate zones. So we weren't able to avoid smoke pollution, and we would expect that we would evolve so that we can better cope with it. Several studies have shown that there is a genetic component to alleviating the risk of pollution exposure, so there must be alleles that offer better protection than others. Therefore there must be some level selection.

If we have been exposed to fire for more than a million years, and there is a selective advantage to cope with it, why are we still dying from pollution? Evolution isn't magic (even if it can seem that way), and it will never be harmless to expose ourselves to pollution. For one thing, it is carcinogenic and will mutate our cells, and that is damaging at such a fundamental level that if it were possible to fix at all, it would have been fixed a long, long time ago. Still, if some people are better, if not perfect, at coping with air pollution, why don't we all have those alleles? One explanation could be that the selective advantage isn't strong enough, or hasn't been in the past, where we should have evolved this pollution resistance. If the life expectancy is very low, and reproduction happens at a young age, then long-term exposure to pollution could be a minor problem; if you do not live past 40, then cancer is not your primary concern. If other factors than smoke cause child mortality, then air pollution is not the main problem either. There will still be a selective advantage, but if other things kill you before the smoke, then it is a smaller fraction of the population that selection has to work on; resistance to smoke only gives you a tiny advantage. Add to that the small population sizes of the past. Selection needs enough copies of the beneficial alleles to really get going, and in small populations, they can easily get lost by chance before they rise in frequency. Now, with a larger population size and more copies of the alleles that cope better with fire, we could see a more substantial selection. However, if we stop cooking over a fire, and stop smoking, then the alleles that could previously be selected for, will become neutral. And we are moving away from cooking over a fire as the world grows wealthier, and more switch to electric stoves, so soon, there might no longer be a fitness advantage in dealing with smoke. Of course, the genes that can protect us from air pollution might also protect us from other pollutants, so they could still be under selection, just not because of wood burning.

The air pollution that we usually think about, smog not smoke, is novel to us as a species. There was no smog until the Industrial Revolution, where burning coal in the densely populated London created the famous 'London Fog.' The composition of pollution has changed from mostly the same smoke that we were exposed to for millennia, to what we now get from burning fossil fuels and from byproducts of industrial manufacturing. So from air pollution, we are exposed to many chemicals entirely new for us. At the same time, the total exposure of the species to this pollution has increased, a consequence of more industry and more urbanization. Rapid and dramatic changes in the air we breathe will no doubt affect us, but for how long? Will we pollute, at this level, a thousand years in the future? I think not. Electric cars, buses, subways, and trams will remove traffic pollution. The electricity production needed to run the electrified traffic, our appliances, and our houses can transition to sun, wind, water, or nuclear power. The industry where we cannot entirely avoid pollution (which I think will be very little of our industry) we can move out of the cities, and maybe even into space. I doubt that we will see the pollution that we have in cities today in one or two centuries; we could have much lower levels of pollution in 50 years if we want to.

There are other sources of pollution than air pollution. We are polluting our drinking water, our oceans, and the ground we walk on, but for these sources, I think the case above can be repeated. We have done much damage to our environment and Earth's ecology, with irredeemable consequences, but when it comes to pollution, I don't think we will continue for long. Today we are already more careful about pollution than we were a generation ago, we are trying to limit it now, and we are slowly eliminating it. The total amount of pollution is increasing with increasing population sizes, and with higher energy consumption in the developing world, but in developed nations, the pollution per capita is decreasing. With sufficient wealth and clean technology, we should be able to eliminate most sources of pollution. Cleaning up the existing pollution will take a long time, especially if we need to wait for groundwater to be pollution-free, but over hundreds of thousands of years, Earth will recover, if we only stop making it worse.

The synthetic material we surround ourselves with, stuff like plastic, is new to our environment. Forget for a minute that it pollutes our oceans—something we ought to avoid and to rectify—plastic and other new materials affect us directly. Materials we touch (or breath or eat) exposes us to chemicals, chemicals that are new to nature, and that we have never adapted to deal with. Some of those chemicals are harmful. That is why we ban them and replace them with alternatives when we can. We are moving away from some of these materials, substances that we know expose us to harmful chemicals, but we replace them with other materials that might also have unintended effects on us. Plastic was invented in 1907, and many synthetic materials are more recent inventions. We haven't discovered what they do to us and how to prevent it yet. We might do that shortly, and then again, we might not. If I guessed on which pollution will stay with us far into the future, it would be those embedded in essential materials, those materials vital for our industry and technology, and not pollution from transport or energy production.

But assuming that pollution is a temporary phenomenon that will only last a few generations more, can it then even affect us evolutionary? Evolution typically operates over tens or hundreds of thousands of years, but it doesn't have to. If the selective pressure is severe, then allele frequencies can change extremely rapidly. Imagine if exposure to a pollutant makes half the population sterile but doesn't affect the other half, and imagine further, that who goes sterile, and who does not, is determined by an allele at 50% frequency. The next generation, that allele will be at 100% frequency—only carriers of the allele can leave any offspring for future generations.

Exposure to pollution will, of course, not have such dramatic effects, if we are even the least bit careful about what we allow our industry to produce. But we are exposed to chemicals today that might still have a similar and dramatic impact on allele frequencies in a very short timeframe. Studies indicate that air pollution might be linked to lower sperm quality and that plastic can release chemicals that act like the hormone estrogen. Estrogen affects sperm count, and sperm count is decreasing worldwide. There is a genetic component to sperm quality. If increasing pollution (for example, exposure to plastic) affects fertility, then pollution can undoubtedly affect our genetic future on a short time scale. Infertile men will not pass their genes, that are sensitive to chemical exposure, on to future generations.

In vitro fertilization, or IVF, is our technological approach to alleviating the consequence of our own pollution, and it can compensate for low sperm quality. If the quality is too low, however, it doesn't suffice unless we add a sperm donor to the equation. Donated sperm must come from a donor with sufficiently high sperm quality, so now we have added an additional source of selection to alleles that add resistance to hormonal pollution. If there are genetic components to the willingness to become a sperm donor—

there might be, although I have never seen a study that examines it—then we would also see a strong selection for these. In either case, while alleles that reduce sperm quality are selected strongly against, the alleles that come through sperm donors will have a strong positive selection. Different countries have laws that limit the number of children that a single donor can be the father of, but international trade of sperm bypass these laws, and a single donor can be the father to vastly more children than he would normally be able to without sperm donation. Sperm donation and variation in fertility give us such a steep selective gradient that we can see an adaptive effect even if the exposure to pollution is temporary.

It is not only male fertility that is affected by pollution. Female fertility is as well, and women exposed to above-average pollution have a lower success rate at pregnancies after IVF. This means that we could have a similar selection through female fertility/infertility. However, women cannot donate eggs to match the sheer volume that men can donate sperm, so the effect will be less extreme.

I am convinced that we can get rid of most sources of pollution by improving our technology. As long as there is a will to do this, we can do so in short order compared to the time scale we consider here. But some pollution could be harder to avoid. If some sorts of pollution remain with us for tens of thousands of years, then it could be something we would need to adapt to, even if their effects on us are minor. If we have a selective gradient and are exposed to a pollutant long enough, then we will get more resistant. Since I have no idea what the future materials might be, I cannot guess at how we would evolve to deal with them, only that we will have to.

Ecological Impact

We are an extraordinarily successful species—you can find humans on all continents of the planet—and the ecological impact per person is arguably higher than any other organism since Cyanobacteria poisoned the atmosphere with oxygen. We are the main suspects for killing most of the paleolithic megafauna as we migrated out of Africa, and only spared some of the African megafauna because they evolved together with us and learned to avoid us. We are fixing that oversight now by destroying their habitats, and I fear that we will have killed off Africa's megafauna as well shortly. We are the cause of a mass extinction. Not just of large animals but a sizable proportion of Earth's species. Unless we get very serious about conservation, and very quickly, our world will lose much of its ecological richness. I am not optimistic that we will act in time to prevent this. I am convinced that we will save many species in zoos or preserves, but many more will join the ones we have already wiped out before we stop our rampage.

We are destroying habitats while we are constructing our own future habitat, our cities. Some species will adapt to urban living. Many already have, like raccoons, rats, doves, crows, and foxes. As cities expand into other animals' natural habitats, more urbanized species will follow. Those species that manage to adapt to cities will have a good chance at surviving and even prosper; the urban environment is not endangered by our actions and will only expand. However, not all animals have this flexibility, and some animals we certainly want to keep out of our cities. If they move close to us, or rather we move close to them, then we will actively try to exterminate them, instead of just killing them as a side effect of some other goal. We will rather exterminate grizzly bears than allow them to roam our streets.

We might slow the extinction process indirectly through increased wealth. The more affluent society is, the more likely that preservation is a concern. It is, of course, a cultural issue, but it requires wealth to make sustainability and preservation a priority; if your worries are survival, then your priorities are elsewhere. If we can make everybody wealthier, perhaps we can make everybody more environmentally conscious as well—a win-win situation.

The main sinners when it comes to destroying habitats, are farming and mining. We burn down forests for farmland, and we strip mine everywhere we find convenient. Improving agriculture technology might help us feed the world while using less farming land. It has done so in the past. Artificial fertilizer is the reason we can supply the current population size with food. Without that innovation, farming isn't efficient enough. Organic farming just doesn't cut it; it requires much more land than modern industrialized agriculture. If we can farm more efficiently, we can leave more land to lay fallow. If we change our diet to use less meat, we reduce the farmland we need even further. If we go vegetarian and do not need to grow crops for food animals at all, we reduce the necessary farmland substantially. If we need less farmland, we can let most of our current fields regress into wild nature. We can add hydroponic and vertical farming to this, to release even more land, and move most of our food production into our cities, so food production only affects our habitat.

Technology might also help reduce ecological destruction

from mining. Not that we can necessarily mine more efficiently in the future, or that we will need fewer mineral or fossil resources, but we are getting better at recycling and producing energy without fossil fuel. If we can get most of our metals and rare-earth out of discarded laptops and smartphones, then we can substantially reduce our need for mining. And with the right technology and processing, 'mining' discarded electronics, rich in the materials we need, will be more economical, so recycling is a win-win. We can do this with technology that we have today; the technology we have in the future is likely to be better. If we get serious about avoiding mining when it can impact habitats, we can take the ultimate route and stop mining on Earth altogether. Near-Earth asteroids are rich in the minerals we need, and with only slightly improved space technology compared to what we have today, asteroid mining is a real option. Because asteroids are so rich in minerals, it might even be economically preferable to mine them instead of mining on Earth. Even if that is not the case, pressure to preserve wild habitats could force mining into space.

We don't need to take up as much living space as we do now, either. Urbanization packs us closer so we, per capita, take up less space. We could fit the entire world's population into Texas if we lived at the same density as current New York City. We would need a few more states for agriculture to feed all the people, though. The point is that if we move closer together, we could free up a large part of the Earth for undisturbed nature reserves. We are not running out of living space anytime soon, there is room both for us and our fellow life, if we move a little closer, even if our population size grows by a few tens of billions.

Nature could benefit immensely from a combination of urbanization, concentrating people at higher densities in smaller areas, and improved farming and mining technology. We need less land for habitation, less land for agriculture, and less mining for resources, relative to our population size. This can free up land that we can then transform into nature reserves. There is the possibility, at least, that urbanization could help us preserve some species diversity to compensate for the destruction we have caused.

We now have the technology for de-extinction, that is, bringing back extinct species. We cannot reanimate all extinct species—sorry, there will not be a Jurassic Park—but if we have an extinct species, that we have obtained its DNA from, and we have a living related species, then it is possible. We take the DNA from the extinct species and DNA from a species related to the one we want to bring back. We modify the living species' DNA to look like the extinct species using a technology called CRISPR/Cas9. Then we insert it into an egg from the host species, where we have removed all the existing DNA. Finally, we let a female from the host species bring the egg to terms. For example, we have the DNA sequence from mammoths, obtained from fossils, and we know that Asian Elephants are a closely related species. So we can go through these steps to use an elephant as a host to bring back mammoths.

We have brought back a species once already, although it wasn't a roaring success. The last bucardo goat, named Celia, died in 2000. Scientists took cells from here and frozen them, and then in 2003 created a clone from them. They created cells to implant in 56 goats, only seven became pregnant, and only one gave birth. The kid died after ten minutes. De-extinction is not trivial, and there is work to be done, but we always get better over time, and we will manage to bring back some species.

There are currently active projects for brining seven species back to life: *Quaggas,* a relative of zebras from South African; *aurochs,* a wild ox from Europe; *Floreana Island giant tortoise*; *passenger pigeons*; *woolly mammoths*; *hearth hens*; and *little bush moa*, despite the name a large (1.3 meters tall) bird from New Zealand. If any of these projects are successful, more will follow.

Bringing back megafauna like mammoths and aurochs might be possible, and would restore ecosystems we destroyed when we caused the extinction of these animals. But if they are simply going extinct again because we have destroyed their habitats, it is a futile exercise. We need to restore their ecosystem in parallel with restoring their species. If we manage to reduce our exploitation of nature, we may be able to restore some of the damage we have done.

I hope that we will have the resources and desire to preserve the world's diversity and preserve endangered species and their habitat in a prosperous future, where we have the resources to do so. If we can stop destroying and start restoring life on Earth, it will have a huge effect on the richness of the world we will inhabit. It will not change ourselves much as a species. We will be just fine in our growing urban habitat either way. But it will be a poorer world we will live in if we do not act soon.

Diminishing Resources

I do not expect that the human population size grows enormously in the near future, despite what selective benefits there might be to having more children. It looks like the current growth is slowing down and will level off as a consequence of the demographic transition we discussed in the previous chapter. We will still see growth until the end of the 21st century, and we will go into the low double digits of billions before the population stops growing. To deal with a world population that is twice as larger as it is today, some things have to change. Especially if the entirety of humanity is getting wealthier and will consume more energy at the same time. We will need to get more efficient with our natural resources, and we need a more sustainable energy production. Otherwise, we will quickly exhaust the world's resources.

The most pressing worry is energy production. Without energy, we cannot heat our houses, and without heat, we cannot survive winters on a large part of the Earth. Without energy, we cannot produce any of our advanced technology. Our industries grind to a halt. We can still grow food, but without modern machines, we cannot do so at the scale needed to feed a population of billions. Our current energy production is predominately based on fossil fuels, and oil and coal will eventually run out.

We will probably have to rely on nuclear power in the short term. We are aiming at reducing our CO_2 emission, and nuclear power is the only stable energy source we have that is both reliable and CO_2 neutral. We have all heard about accidents with enormous consequences, Chernobyl comes to mind, but it is the same as when we hear about plane crashes. They are rare but more dramatic. When a plane crashes, it is breaking news for several days, while a car crash is rarely in the news at all. However, flying is still safer than driving. A nuclear accident is a catastrophe, but there have only been three major accidents in the almost 70 years that we have had nuclear power—Three Mile Island (1979), Chernobyl (1989), and Fukushima (2011).[2] The death toll directly attributed to these three accidents is 30, and they are all from Chernobyl. The radioactivity leaked by them might have caused an excess number of cancer cases in the low thousands on top of that. Coal power plants kill hundreds of thousands each year through pollution— some of this pollution also happens to be radioactive. In comparison, nuclear power is extremely safe. For each terawatt-hour produced, coal kills 24.63 people, oil 18.43, and nuclear energy 0.074. Nuclear energy is much safer, *and* it is CO_2 neutral, unlike the other two.

Renewable energy sources are even safer, with 0.035 deaths per terawatt-hour for solar energy, 0.024 for hydropower, and 0.019 for wind. Most developed countries are converting their energy production to renewable energy, and it is a better solution than nuclear, except that they are less reliable. The energy output depends critically on the

---

[2]There were no fatalities from the Three Mile Island accident. Two were killed during the initial Chernobyl accident and 28 from radiation poisoning in the following months. It is estimated that 4000 deaths from cancer would not have had cancer if it weren't for the increased radiation. There were no fatalities from the Fukushima accident, but 34 died in the evacuation panic. In comparison, 19,000 died from the tsunami that caused the accident. Estimates of excess cancer victims vary from zero to a few hundred. They are uncertain because the excess radiation after the accident is very low—and we mostly do not know how very low increases in radiation affect cancer risks.

weather, which we do not yet control. We do not currently have efficient technology for saving up energy when it is sunny or windy such that it is available whenever we need it. We will get there, I am sure, but until then, renewable energy sources must be supplemented with a reliable energy source, like coal, oil, or (preferably) nuclear energy.

Another issue with renewable energy is how much land we need. Windmill or solar parks take up a lot of space, and when our energy consumption increases, this is an issue. Our energy consumption will rise, both because there will be more of us, but also because our technology uses more energy. All our tools are getting more energy-efficient, but we compensate for this by getting more devices, putting IT everywhere, and expect it to do more every year. If we want to save the Earth's ecology as well as its atmosphere, we cannot cover the world with windmills solar cells. A wind farm requires up to 360 times as much land area as a nuclear plant for the same amount of energy produced. A solar photovoltaic (PV) facility 75 times as much land.

In the short term, we are better off relying on nuclear fission technology but probably supplemented by renewable energy sources. It is a well-tested technology, it is safe, and it doesn't pollute the air. There are radioactive waste products, but this problem is far less than many believe. Most of the waste from a nuclear power plant needs to be stored for a measly few decades before it is no longer a health risk, not hundreds of thousands of years. With modern reactors, only around 3% of the waste needs to be stored for thousands of years, and tossing that into mine shafts is a small price to pay for stable, safe, and (otherwise) non-polluting electricity. It will probably be a large part of our energy production for a while, at least until uranium runs out.

In the longer term, we can look to space. With technology within reach today, we could build power satellites that harvest sunlight (it doesn't get cloudy in space) and beam it down to Earth as microwaves. We would need a large fleet of such power satellites, but it is within our technological capability to do this. And at some point, when fossil and nuclear fuels run low, it could be necessary. Then there is also fusion. Fusion power has been just around the corner since the 1940s and might remain around the corner forever. However, if we ever get it, we do not have to worry about energy ever again. It would essentially give us all the free and non-polluting energy we could ever want.

Electricity and heat production aren't everything. For transportation, for example, we have more work to do; we still don't have the technology to go completely electric. We can run cars and public transportation in cities with electricity, but we need better batteries and faster charging to make electric vehicles attractive for long-distance drives. We need much better batteries for trucks, and we are far away from the kinds of batteries that can run supertankers and long-distance flights. But even if it takes a century or two to get there, I am optimistic that we eventually will.

Once our energy production is clean, the problem I listed earlier with air pollution goes away. So do the selective pressures they created, so we might only have a short time to see any adaptation there. It doesn't mean that all types of pollution disappear, of course. We still have the materials we produce that might be polluting and adversely affect us. If we stop burning fossil fuels, we have more of the organic compounds we use for making plastic, for example. And plastic is essential for most of our technology at present. If we don't burn oil, it will stretch how long we can use plastic in the future, but we can hopefully alleviate many of the ill effects plastic has. Still, we might be able to, and want to, produce plastic for a while, until our oil does run out.

We call oil nonrenewable, but technically it is renewable. You take organic material and put it under pressure underground for millions of years. We can make oil, but we have to be patient, and patience is not something that our species is particularly known for. We will run out of oil, and we need to replace plastic with other materials. I don't know what those materials could be, that is up to future scientists and engineers. But I do know that if they have toxic effects, like plastic, for example, they could affect fertility or increase the risk of cancers. Then we will see selective effects like those I have described earlier in the chapter.

Over the next millennium, we will probably need to change our food production. If we double our worldwide population size, and everyone wants to move to a diet close to what we now have in the developed world, our current food production is not sustainable. Massive industrialized meat production—animal welfare aside—still requires a lot of farmland. If, for nothing else, and even if you keep the animals locked up inside, you need to grow feed for them. There is still a lot of land to cultivate, but this will have huge impacts on our ecology and diversity, so we might want to avoid that. Then again, we might prioritize ourselves over nature, but I do not think we will continue to do this. We will likely reduce our meat production substantially, in as little as a few decades, and what meat we still consume could be synthetic (built from animal cells but not actual animals). Likely, though, most would change their diet to be entirely plant-based.

A change in diet can create a selection gradient. We evolved to be omnivores, and this puts some requirements on our diet. There are so-called essential nutrients, which are molecules and minerals that our body cannot produce, at sufficient quantity or at all, so we have to get them through our diet. They are called essentials because our bodies cannot produce enough of them, and not just because we need them to live. We do need them, we would die without them, but there are lots and lots of things we need to stay alive. Only those we cannot produce ourselves are called essentials. Don't blame me; I didn't come up with the term.

For practically all of our essential nutrients, we are talking about compounds that we have needed throughout our entire evolutionary history. They are fundamental to how our cells and bodies work, and practically all life on Earth needs them. We evolved the need for the rest of these molecules and minerals later, but still very early in our evolutionary history, and none are uniquely needed by us. All animals need the same minerals and molecules that we do. They are just not 'essential' for all, in the technical sense, where essential means that we cannot produce them ourselves. We need the same stuff, but there is variation among animals in which of the molecules different species can produce.

Our ancient ancestral species could produce many of our essential nutrients themselves, but we lost the ability over a long evolutionary history. We can lose the capability to construct something if we can get it some other way more efficiently. If we can get the molecules and minerals through our diet, then our body does not need to produce them, and the genes we had for producing them can atrophy. Once these genes are gone, we are stuck with the need to get the essentials through our food. It is easy for random mutations to destroy a gene, but it is tough to fix it again. It didn't take much to break Humpty Dumpty, but it was pretty hard to put him back together again. Not even all the king's horses and all the king's men could do it. Once we have lost the ability to produce a molecule that is vital to us, and we can only get it from our food, there are constraints to how much we can change our diet. We must get our essential nutrients if we want to survive, so we must eat foodstuff that contains them—there is no way around it.

The diet we evolved to eat includes meat, and there are essentials we cannot readily get from an exclusively plant-based diet. With supplements, this is not a problem—this is why there are still vegetarians around—but without supplements, there is a risk of deficiency for one or more essentials. With a varied diet that includes meat and dairy products, there is rarely a need for supplements; we get what we need from the food we evolved to eat. It doesn't mean that we cannot live without meat unless we take supplements. We can get everything we need without animal products, even with a vegan diet, at least if you add mushrooms and algae to your food source. Still, supplements are usually recommended, because otherwise, you have to be very careful with how you compose your diet to ensure that you get all the nutrients you need. The diet we evolved with, gives us what we need, or we wouldn't be around anymore. However, if you change your diet, it is up to you to ensure that you still get everything you need from your food. If vegetarians and vegans manage their food intake correctly, or if they take supplements, there are no drawbacks to their diet. They are as healthy as people who still eat meat.

If you switch to a vegetarian diet, you will probably go for the supplements option. It is arduous work to compose your food intake, so you get all the right essential nutrients, and if you reject supplements, you do need to do this difficult work. However, there is a risk with supplements: excess intake. It is unlikely that you get, for example, vitamins at a toxic level from your diet. Again, because we evolved to eat this diet so our bodies can handle the levels of nutrients we get from it. But it can happen with supplements. If we can get our body to produce or extract what we need from a new diet, and no more, then we would be better off. There might be a selective advantage to this, getting better at extracting whatever you need from your diet.

For most molecules that we need, we have genes that produce them. For essential nutrients, we might still have genes that produce them, but the molecules are essential because we do not produce them efficiently or at high enough quantity. If we have lost a gene completely, it is practically impossible to get it back. However, if we have a gene that simply doesn't produce the nutrient efficiently enough, we already have the capability we need; all that selection has to do is dial up the nutrient production. DHA is an omega-3 fatty acid that is vital for brain development (which we would certainly expect to be essential for fitness). We can get it from fish, for example, and you get plenty of it if you eat fish regularly. To produce it by ourselves, our body needs to do some extra work. We can construct DHA from another fatty acid, ALA, and we can get ALA from some plant food, such as chia seeds. This ALA to DHA conversion is inefficient, though, and vegetarians have lower DHA levels unless they take supplements. With a plant-only diet, I could imagine selection for something like improving the efficiency of the ALA to DHA conversion.

For essentials that we cannot produce at all, we must necessarily get them entirely from our diet, and we cannot just turn up our own body's production of them. Still, there might yet be room for selection and evolution. We have a metabolism for extracting these essentials from our food, and this metabolism is optimized for the typical food we eat. But we can often obtain the same nutrients from different food sources, just more or less efficiently. If we are optimized to extract the nutrients from one food source and switch to another that we are less efficient at metabolizing, we once again risk a nutrient deficiency. But your body could evolve to be better at extracting the nutrient from the alternative food source.

You cannot produce iron, and we need iron to stay alive. It is a crucial component in hemoglobin that transports oxygen around your body—no iron, no hemoglobin, no oxygen transportation, no you. Iron is an element, and we cannot make those (that is the sun's and supernovas' job). So we must get our iron through our diet. There are two kinds of iron that we can get from our food. One, heme iron, you can only get through meat and the other, non-heme, through both meat and plants. Meat contains 40% heme and 60% non-heme iron, while plants only have non-heme iron. The problem with a plant-only diet is that non-heme iron is a lot harder for us to absorb. You get enough iron if you eat enough iron-rich plants, lentils, chickpeas, beans, quinoa, and such, but once again, you have to be conscious about how you combine your food when you eat differently from how you evolved to eat. If we go vegetarian, there might be a selection for something like getting better at absorbing non-heme iron.

Evolving to a different diet doesn't have to be our destiny, though. We have supplements, and as long as we take them, we don't have to worry about how well we absorb our nutrients from our food. But if we decide to stick with supplements far into the future, and toxicity is an issue, then adaptation to avoid this problem is a possibility. A more straightforward solution, however, is to change which nutrients are in our food. We change what we eat, and at the same time, we modify the food in our new diet, so it contains the nutrients we need. This way, our new diet gives us precisely what we need in the quantity that we need.

Rather than explicitly taking supplements, we can add them to our food stables. Then we don't have to worry about it ourselves. In many places, vitamin D is already added to milk, and I don't see why we can't add all the necessary nutrients to our chosen diet. Of course, as a species, we tend to adjust the environment to us rather than adapt to our environment, and we can readily consider our diet is part of our environment. After all, it is. We can modify the plants we eat so they can provide us with the nutrients we need. For all our essential nutrients, some organisms can make them or extract them from the environment and make them available to us. If it weren't so, we would already be dead. We can only eat organic matter, and some form of life must have created our essentials or extracted them from inorganic matter, such that we can eat them. We can take the necessary genes from these organisms and use them to modify our food stables and base our diet on genetically modified organisms (GMOs).

Some essential nutrients will be natural to get from GMOs. Golden rice, for example, is an existing GMO that produces vitamin A. Vitamin A deficiency kills an estimated 1–2 million each year and causes blindness in 250,000–500,000 children. Golden rice was developed to alleviate this. It naturally ran into resistance from anti-GMO groups that don't want us to modify plants. We have, of course, modified our food sources for tens of thousands of years, and no food stable evolved naturally. None. They are all the results of selective breeding. The only new thing about genetically modified organisms is that we can do it faster and have more control over the outcome. We are not building new genes at random, but moving existing genes from one organism to another. We need to be careful with this, of course, because everything we do have consequences, but writing off GMOs entirely because of fear, dooms many people to misery. It is Luddism at its worst. We can use GMOs for good, and we should, and we probably have to in the future. I am optimistic that once a technology is a few hundreds of years old, it is 'natural' and no longer playing God, so eventually, no one will give GMOs a second thought.

With GMOs, we can add genes that produce our essential nutrients to our food stables. Some nutrients might be easy to add, like vitamin A in golden rice. Some pathways for essential nutrients might be much harder to add to an organism. With time, improved technology, and sufficient motivation, I believe that we can do it. If we decide to go this route, then the selection for adaptions to a new diet will mostly disappear again. An artificially created diet has to be almost perfect, though, to remove selection altogether. Before we get there, we would stop improving our GMOs when they are good enough for our dietary needs. Maybe after that, we would improve on taste, looks, or ease of preparation, but I doubt we would keep optimizing the mineral and molecular composition. As long as there is a selective gradient for how well we deal with our diet, nature will keep optimizing us.

Climate Change

The most significant way we are currently changing our environment is through emissions that alter our atmosphere. We have emitted CO_2 since we started burning wood, but since we really got started burning fossil fuel with the Industrial Revolution, we have accelerated our emissions to the point where we affect the entire climate system.

We might be able to alleviate climate changes—although we have been reluctant to do much about them so far. I am optimistic that we can take action to slow, if not reverse, our effect on the climate. I think the reluctance to do so is only the inconvenience of what it takes and not our ability. We don't want to give up power plants, planes, and automobiles. Even the dignitaries that attend climate change summits arrive by private jets. There were 1,500 private jets at the 2019 Davos summit. They couldn't even be bothered to jetpool. The hypocrisy is so thick that it must be affecting the atmosphere all by itself.

I don't think we can stop climate change entirely by using less energy. The dramatic drop in living standards in the developed world will prevent people there from reducing their emissions if that is the only option they are given. And the developing world is not going to sit and watch the living standard in the developed world, know that they can get there as well if they only increase their energy production, and then decide not to do it, because that will save the planet. They are not going to take one for the team. Especially considering what the team has done to them in the past. I think the only realistic way to reduce CO_2 emission levels is through better technology that replaces the polluting alternatives we rely on today. Electric cars—if we can manufacture them using fewer resources than we save by driving them—is one step in that direction, provided we can create the electricity for them without fossil fuel.

We already have wind, solar, and nuclear power today, so we know how to create CO_2 neutral energy. We might have fusion power in the future or a combined power grid that makes wind and solar a practical alternative. With renewable electricity, we will go far. People are working on electric planes as well. These might not be able to go long distances, but it all helps. High-speed trains can handle some of the traffic, as long as we are not crossing oceans. And maybe even then, if we get the necessary technology to drill tunnels that long. We have video meeting technology, so business meetings and conferences could be handled without travel. The COVID-19 crisis has shown us that this is doable. And if people are serious about the climate, they should only attend climate summits electronically. Considering the private jets I mentioned above, I don't think they will, but they should. We should all do our best to reduce our climate footprint, and I think the best way to do this is through better technology.

But the survival of the species is not in danger from climate changes, so we probably shouldn't use that argument to convince people to take it seriously. I do not believe climate changes or smaller ecological diversity will have a sizable effect on our future evolution. We are turning into an urban species and do not need diverse habitats. Even with the worst predictions of rising temperatures and sea levels, there will be plenty of places for us to live. Evolutionary speaking, there is no reason to consider climate change an existential risk for humanity.

I do not want to belittle the problem; it is certainly one we should take very seriously and work to alleviate. Please don't misconstrue what I say. We should take the warnings to heart. The consequences that climate scientists predict are very likely real. We should do all we can to mitigate the damage we have done. Measured in human suffering and death, we are looking at a disaster. Climate changes will cause widespread starvation. A large part of humanity will need to migrate. The economic effects will be massive if that is your primary concern. So we need to act and act soon. If we could go back in time and act sooner, we should do that. But don't worry about extinction; our species has already dealt with substantial climate changes, including ice ages, over the 200,000–300,000 years we have existed, and over the millions of years, our ancestral species were around. They adapted and survived, and they didn't have the technology that we have today. We will be fine.

Rising sea levels will force some to relocate. It will affect the availability of food if farmland becomes unusable. But beyond what I have described earlier in this chapter, I doubt that climate change will subject us to extreme evolution. None of us will grow fins; this isn't Waterworld. Life could get rough for a millennium or two, but we will adapt to a different climate over the next million years.

If climate change means massive migration, what it probably does, we end up mixing populations to a larger degree than I have described earlier. We will lose cultural and genetic diversity faster, but the end station is the same. If climate changes destroy our agriculture, which they could, then we must change our diet to compensate in ways I have also discussed. But it is just speeding up trends that we will enviably see anyway. Climate changes will affect the ecology, and we will lose species this way, but this is only an acceleration of what we are already doing by destroying habitats. We humans, however, will survive. I am not worried that climate change will have a significant impact on our future evolution. Nor do I think that loss of species diversity will impact us much. Still, it is something I greatly wish that we would strive to avoid.

But what if? What if climate changes bring us close to extinction? The human population could go through a massive bottleneck, as we considered when we discussed global pandemics a few chapters back. We could lose most of our genetic diversity, and allele frequencies would shift dramatically. Not all populations will be hit equally. Some regions are more fragile to climate change than others, and if the disaster strikes unevenly, we will see a genetic shift based on who gets hit the most and who gets hit the least.

But our misery will not last forever. We will learn to live in a changed world. It may take centuries or even millennia, but we will learn how to produce enough food. We will build energy production plants, build dwellings and cities that can sustain extreme weather, and get our civilization back. We will recover, and we will start multiplying and expanding again.

Those who make it out on the other side of the crisis could carry more genetic defects because selection against them would be weaker during the bottleneck. Once on the other side, however, I suspect that we would see an exponential population growth, as fast as our new resource production can handle. We would quickly bounce back to a large population size and back to an effective selection and fast evolution. The crisis might have shifted our genes, but not changed the trajectory of our future. Merely delayed it for a bit.

CHAPTER **6**

Going Forward

Growing older and adapting to modern-day diseases are relatively short-term effects of our changing lifestyle. We have largely already transitioned to this new world. The changes I have described in the previous chapters are already happening, and all I have written about our possible future evolution are consequences of what happens *now*, not in a thousand or a hundred thousand years.

In this chapter, I will try to imagine what could happen a little further down the line. I do not have the imagination to predict the technology or culture we will have in a million years. Still, I will venture some guesses at what future technology *could* look like, and how we will evolve as a consequence. I will only speculate on what could happen with technology in the next couple of millennia, and then extrapolate what evolutionary consequences this could have for the next million years—starting with the least dramatic changes to humanity and finishing with a humanity entirely different from what we are now.

Along the way, we will leave natural selection and natural evolution. It might be paradoxical that I do this in a book about our evolution, but I don't think our future will be guided by natural selection and biological evolution in the way it has in the past. We have the technology to direct our own evolution, and I firmly believe that we will.

I am convinced that we will start editing our genes to improve ourselves within no more than a century. We have the technology now, and it is a question of time before we use it. It won't stop us from evolving, but it will be a very different kind of evolution. We might even leave biology behind us entirely and become machines. I do not dare to guess how likely this is, but I expect that we will have the technology to make this possible within the next thousand years. It won't turn off evolution, I do not think that this would be possible at all, but we, and not nature, will do the selecting.

But we start more down-to-earth, or rather, under the sea.

Colonizing the Oceans

I have generally assumed that global overpopulation is not going to happen, because our numbers will not continue growing. The human population is *currently* growing at an alarming rate, but this won't last. In the 20th century, we went from 1.69 billion in 1900 to six billion in 1999 (we were at two billion in 1927, three in 1960, four in 1975, five in 1986, and six in 1999; we have to go back to 1803 to find when there were only one billion people). We reached seven billion in 2011, and we expect to reach eight in 2024. But it is not the exponential rate we have discussed earlier in the book. It took 124 years to go from one to two billion people (1803–1927), 33 years to go from two to three billion (1927–1960), 15 years to go from three to four (1960–1975), 12 years for four to five (1975–1987), 12 years for going from five to six (1987–1999), 12 years for going from six to seven (1999–2011), and it will be 13 years to get to eight (2011–2024). So the intervals between when we grow by one billion are 124, 32, 15, 12, 12, 12, and 13 years. With exponential growth, the time it takes to increase a number by a fixed amount, say a billion, gets shorter and shorter. We see that for the first four numbers. Decreasing numbers here doesn't necessarily mean that the growth at the beginning of the 20th century was exponential. But the shrinking intervals indicate that the growth was going increasingly faster. The growth isn't speeding up after that. We have a linear growth if the time it takes to increase by one billion is the same when we go from five to six as it is to go from six to seven and seven to eight. With linear growth, the population is still growing, but the growth is neither speeding up or slowing down.

Of course, if the population is still growing, the difference between exponential growth and linear growth is just a question of how long we have to wait before overpopulation causes our civilization to collapse. With exponential growth, it happens incredibly fast, while we can wait many millennia with linear growth. But we do not even have a linear growth. Looking at intervals between increasing billions in the way that we just did is a crude analysis. If you look at the growth per year, you will find that the rate of growth in the early 20th century peaked at 2.2% in 1962 and 1963, and it has been decreasing since. Our population size does not grow faster and faster; instead, it is slowing. The growth only looks linear now, because we are at the point where the growth is changing from increasing over time to decreasing. UN projections are that it will take 13 years to go to 8 billion (2011–2024), 14 years further to go to 9 billion (2024–2038), 18 years to 10 billion (2038–2056), and 32 years to go to 11 billion (2056–2088). So 13, 14, 18, and 32. The intervals are increasing. The population is still growing, slower and slower, which means that we might see the growth stop before long (say in a century or two). The latest UN World Population Prospects (2019) estimate that the population growth will level off, and the world population will stop growing, around the end of this century, topping at around 11 billion. The exact time for when the population growth stops, or the size of the population when the growth levels off, doesn't matter much for the discussions in this book, only that the population size doesn't continue growing uncontrollably.

But in the demographics chapter, we talked about the selective advantage of having more children. If there are no stronger reasons against having many children, then selection will work its magic to increase the population growth again. What could happen then? Eventually, the world's population would have to introduce some sort of birth control, a new form of China's one-child policy, or something even more draconian, but let us consider what could happen before we get to this point.

If the human population keeps growing, it means that cities will keep growing as well. Cities will turn into megacities, and cities will start to merge as they expand. The agglomeration of cities will limit the area they can expand into. The building material we have available might limit how far we can grow in height, so we cannot merely build higher skyscrapers so we can pack more people into the same area. The building materials need to compete with gravity, after all. We can expand some distance underground by living in basements. We can dig deeper basements than we can build skyscrapers. But we can only go so deep before these basements become an engineering nightmare. With more people, we will need more farmland to feed the population, although changing our diet and our food production might reduce the land necessary for agriculture. We can pack our food production into our cities if we turn to a vegetarian diet and vertical (hydroponic) farming. This could free up more space for cities that we can pack people into. Still, as the population keeps growing, we will need more and more space. We might initially want to preserve natural habits and avoid species extinction, but we do not have a good track record when it comes to setting nature's needs over our own desires. We will not accept being packed together in denser and denser cities, with smaller and smaller living quarters, simply to preserve a bit of nature, even if it could help protect a large part of our planet's diversity. We can always rationalize taking just a little more of the available land until there is nothing left. Cities will grow, and if our population growth continues unabated, they will eventually cover all available ground, from pole to pole and around the equator. Cities can grow vastly larger before this is an issue, but we are considering an exponentially growing population over centuries and millennia. Eventually, we will run out of surface area.

If cities expand to cover all land, and still need to grow as our numbers multiply, we will be forced to expand in some other direction. Obvious directions are the sea and space.

We consider the sea here; I will get to space in the next section. Compared to the Earth's landmass, the sea is vast. In surface area, the sea is more than twice as large as the land, and there is much more room in the volume under the surface. We can place habitats at any depth as long as they can resist the water pressure—which practically means everywhere if we build the habitats sturdy enough. We do not have to build towers or dig mines to expand vertically, as we have to on land. We can build habitats at any level. The water assists us in fighting gravity, and the same gravity lets us go deep without digging. We can fill the volume of the sea.

The first place we will move is onto oil rigs and ships. There is no vast difference between living in cities or living in a similar environment on the surface of the sea, so there is no difference in the adaptations we will undergo there. But since the volume of living space is much larger under the sea, we will eventually move there, and there might be a cause for adaptation for living underwater.

While living on the ocean likely means living in conditions little different from those in the future cities, early underwater settlements will certainly mean smaller living quarters. The psychological effects of living in cramped quarters and quarters shared by many other people are not entirely clear. Simulated Mars missions show that these conditions can cause stress, depression, insomnia, and hostility. Some can handle these conditions better than others, and if this is partly genetically determined, we should see adaptations. Mental instability will not look attractive to mates, so it will naturally be selected against. We won't see as great a selection as we would see in larger populations, though. Selection is weak in small populations, where allele frequencies predominantly change in stochastic manners. But considering that there is vastly more living space in the oceans than on land, if we are going through exponential population growth, a large part of humanity, and perhaps the majority, will move underwater. Once the population size is large, selection against the detrimental psychological effects of living in underwater habitats can accelerate.

If there is plenty of gene flow between surface dwellers and the ocean dwellers, the selection for living underwater will initially depend on how adaptive the traits needed underwater also are above the sea. An allele that is adaptive under the water but not on the surface will not rise in frequency if the gene pool is repeatedly flooded by genes from the surface, where the allele doesn't rise in frequency because it isn't under selection. Here, one could argue that cramped quarters in cities are similar to quarters underwater, so the relevant alleles would be adaptive there as well, the more urbanized we become. Cramped urban living can also have negative psychological consequences, and adapting to the sea or dense population urban areas might overlap. The selection under the sea will just be stronger than above water. Unlike in cities, underwater living likely means constant contact with the same small number of people, and you cannot just go for a walk to get some alone time. So the conditions are more extreme there than in the above-water cities. If there is strong selection underwater and weak on land, we will still see adaptation, but slower than if a larger population lives in the ocean. The larger the population living in the ocean is, the stronger we should see the selection for alleles that are beneficial there. Thus, if we see a steady migration into the sea as the human population grows, we could see an accelerating adaptation.

Our psychology is a complex system of interwoven desires, drives, fears, and wants. Selecting for some personality traits will also affect others. How our psychology will change when we move into the ocean is hard to say, but it could be profound.

Another issue with living underwater is vitamin D deficiency. We can get a lot of vitamin D from fish, and fish might still be readily available in the ocean even at this point. But our primary source of vitamin D is sunlight, which will be in short supply deep underwater. At high latitudes, humans have developed lighter skin that makes vitamin D production more efficient (at the higher risk of skin cancer, though), but this will not be an option if natural sunlight is not available. Getting more efficient at synthesizing vitamin D from sunlight has no effect if there *is* no sunlight at all. That being said, getting vitamin D from food supplements is not a major issue. In many places, this already happens, especially at high latitudes where sunlight is scarce during the winter. A civilization that can maintain underwater cities will not have issues with producing supplements to alleviate the adverse effects that the lack of sunlight will have on vitamin D production.

Lack of sunlight, however, also has a psychological impact. Winter depression at high latitudes is common. Light therapy has some effect on winter depressions, but there is room for adaptation for coping with less sunlight. It is only a minority of people who suffer from winter depression, and if depression is sufficiently debilitating, then susceptibility to it would be strongly selected against.

Colonizing Space

The oceans are vast, compared to the surface area of the Earth, but the living space there pales compared to the volume available in outer space. We cannot ship an entire growing population into space—we breed faster than we can load people into rockets. However, if we reach the point where the Earth is overpopulated, and we have to take draconic measures to stop population growth, some may prefer to leave the planet to start a new life in space, where living space and resources are virtually limitless. We won't fill up the sea for hundreds of thousands of years, and by then, we will certainly have the technology to build space stations and colonize other planets. Some colonists will already live out there and will have paved the way. If we reach the point where the Earth is full, wave after wave of colonists will follow these pioneers. Growth must by necessity stop on Earth, but it will continue in space, and over a million years, the off-world population could dwarf that of Earth.

Many people today dream of colonizing space. Jeff Bezos wants to build Moon bases, Elon Musk wants to start a Mars colony, and they are not the only ones with these visions. The first space race was a power play between two superpowers. If space colonization happens today, it is driven by a passion for colonization and a chase for profit. Both are strong motivators, and less likely to fade away as fast as the Cold War Space Race did when 'peace broke out.' Passion will provide the visions, but passion alone won't get you there; someone has to pay the bill. Luckily, there are profits to be made in space. Currently, only in sending cargo into orbit, such as satellites or material to space stations, but as we discussed in the last chapter, asteroids are rich in minerals. As resources on Earth diminish, or we decide not to destroy ecosystems to get the minerals from there, asteroid mining could be a profitable business. We don't need to go to the asteroid belt to get them either. There are plenty of near-Earth asteroids we could mine, to bring resources to Earth. The technology we need to develop to mine asteroids will enable us to expand further into space. We can use the material we mine to build space stations far cheaper than if we had to send the material into orbit from Earth. We can mine asteroids near the Earth to construct crafts for exploring and exploiting resources further out in the solar system.

The private-sector space industry is only an option for the extremely wealthy today. But the world is getting richer, and improving technology such as reusable rockets is making space flight cheaper. Over time, space opens up for second-tier billionaires, and there will be increasingly more billionaires. Not all of them will have a passion for space exploration, of course, but the number that does will increase. I can imagine that these space tycoons will enter a long era of competition, both over profit and prestige. The fight over prestige is where I believe we will get our first space colonies—free-floating in space, on moons, and on planets. Once space flight is cheap enough, colonists will pay to move to a new life in space.

We can go to the Moon with the technology we have today. The first Moon landing was more than 50 years ago, July 20, 1969. We stopped going there almost immediately, with the last landing on December 7, 1972. No one has walked on the Moon in my lifetime. We stopped going because going to the Moon then was never about colonization but about showing strength during the Cold War. If we wanted to, we could go there to build a permanent base, and that could grow into a full colony. We could do it today, and the main reason we haven't done so yet is the cost. When going to the Moon as an (indirect) military project, funding was not a problem. As a scientific or colonization project, it is too costly—at least until now. Now the world is getting wealthier, and that is gradually making the Moon an easier target. Several countries, and even some private citizens, have already announced plans to return to the Moon, and some of them wish to build bases there. If governments do not colonize the Moon, then private enterprise will. The Moon is reachable by private companies—if not today, then within the next decade. And they might go, if not for altruistic motives, then to sell trips for space tourists. Both SpaceX and Blue Origin have plans to build a Moon base in the 2020s. Once there is a base there, I would expect a permanent settlement to follow shortly.

To go further into space, we need to figure out how to deal with radiation and prolonged stays in low- or microgravity. On Earth, our atmosphere protects us from most of the solar and cosmic radiation that permeates space, but outside the atmosphere's protection, we are hit by the full dose. This radiation won't kill you immediately, as should be obvious considering the number of astronauts, cosmonauts, and taikonauts who have survived trips to space. But every day you are in space instead of safely down here on Earth increases your risk of cancer. It is not a substantial risk to spend a few days or months in space, but if you live there for your entire life, the risk is considerable. An even greater concern is the low gravity. Astronauts suffer from bone loss (1–2% per month) and their muscles atrophy despite rigors exercising regimes. We did not evolve to deal with the environment of space, and it is deadly to us. However, with the technology we need to have before we can move to space, we do not need to change ourselves too much. We can build space vehicles and habitats that alleviate the problems.

We do not need new science to handle radiation and microgravity, although future technology might make them more manageable. It is mostly an engineering problem. We can shield people from radiation if we bring enough material for a shield. Water would work, and we need water on a space trip anyway. Just don't drink all of it, or build your shield out of urine if you do. We can create artificial gravity by rotating our space ship. The centrifugal force will push us out against the hull and be indistinguishable from normal gravity. We haven't tried building such spacecrafts yet, but there is nothing that should prevent us from doing so with sufficient motivation and funds. If we construct a spacecraft with shielding and artificial gravity, we have also built a space station that can work as a permanent base in space. The International Space Station is not a place you can live—the radiation and lack of gravity will eventually get you. However, if we build a craft that solves the radiation and gravity problems, we do not need to leave it to live on Mars or the Moon. We can just stay in space. We probably do not want to, if the craft is small, but building larger space stations once we reach this point is, again, just an engineering problem. It is within our grasp to colonize the Moon and Mars. Venus, if we can figure out how to build a colony high in the atmosphere (we certainly do not have the technology to build a base on its surface). We can fill the inner solar system and the asteroid belt with space stations.

If we want to go even further out in space, the problem is energy. In the inner solar system, we can get the energy we want from sunlight. In the outer solar system, we need another source. Probes that we send into the outer solar system today are powered by the decay of radioactive isotopes. We probably need nuclear power as well for outer system colonies. Today, sending a nuclear power plant into space is a no starter, but who knows about the future? Of course, if we get fusion power, then energy is practically free and readily available. Let's assume we get fusion, then we can comfortably settle amongst the outer planets. If we can live in the outer solar system, we can also move even further. The Oort cloud, a cloud of mostly icy rocks on the extreme edge of the solar system, is vast, possibly stretching one or two light-years out from the sun. If colonies spread out through the Oort cloud, they will be halfway to the nearest star, Proxima Centauri, when they reach the edge of the cloud. If Proxima Centauri has an Oort cloud as well—and why shouldn't it?—then the end of the sun's and the beginning of Proxima Centauri's cloud will blend into each other, and colonies can seamlessly diffuse through both clouds. Then we will have colonized another star.

Long before we get to the point where diffusion of colonies gets us to Proxima, our technology and experience with building and sustaining space stations will enable us to go to other stars—including stars more distant than Proxima Centauri. I will not assume that we have faster than light travel; nothing in current science suggests that this is an option. However, if we can live indefinitely in space, we should, with a little more work, also be able to build large generation-ships—ships where multiple generations live and die during its journey. With a big ship and a sufficiently large population on it, we should be able to travel to other stars. Those that start the journey might never see the destination, but their descendants will. If the starcraft is constructed like a space station, where generations are born, live, and die, then the only difference between a generation-ship and a space station in the solar system is the longer communication delay. If the population is large enough on the space ship, slower communication with the solar system will not be a problem.

Then again, if our lifespan has increased enough by then, we might survive centuries of travel and make it to at least the closest stars within a human lifetime. Once we can get from one star to another and have the technology to extract resources from asteroids and planets in a new solar system, we can continue our expansion indefinitely. We arrive in a new system, settle it, build new space stations, and perhaps planetary colonies, and eventually, we will build new interstellar ships that take us to even more distant stars. If our spaceships can reach one-tenth of the speed of light—which may or may not be possible with the technology we will have at this point—we would be able to colonize the entire galaxy—which is about 100,000 light-years across—in one million years.

What adaptive forces could be at play when we start living in space? I find it unlikely that we could evolve a defense against radiation—it damages our DNA at such a fundamental level that it is hard to conceive. We might get more resistant to various kinds of cancers, but we would be able to build sufficient shielding for space crafts at a substantially faster pace than evolution could adapt us to live in space. My guess is that the same is the case for evolving to deal with microgravity. It might be evolutionarily possible, it likely is, but since it is a small technical challenge to build crafts that create their own artificial gravity by rotating, there won't be a selective gradient long enough for this to matter.

I could imagine selection acting in another direction, though, one that we have discussed before. If there are practically unlimited resources, and with improved automation and fusion power—which I would expect us to have once we start spreading out amongst the stars— there is no penalty to having huge families. Such technology will also be available on planets, of course, but not the unlimited space to expand into, so the family size on the ground would be limited while it will not be in space. If this is the case, then genes that push people to have more children will spread, and it is these genes that will colonize the galaxy. This could be genes that affect the choice for having more children, which would have an immediate effect, but also genes that let people have children earlier and later in life, for example, by starting puberty earlier and delaying menopause. Maybe genes for larger litter sizes—twins, triplets, or more—could become the norm instead of the exception. As we spread across the Milky Way, those that reproduce faster will outcompete those that reproduce slower and become the future of the species.

Becoming New Species

There is more theory than concrete knowledge about how species form. We know that populations will inevitably change over a sufficiently long time, and there will be incompatibilities that prevent interbreeding when there are enough changes in the genome. At first, the results of interbreeding, so-called hybrids, will be viable but have a selective disadvantage. We can see this in nature when previously isolated populations of the same species come into contact. The next stage is that hybrids are still possible but sterile. An example is mules that are horse-donkey hybrids. At this stage, it is impossible to transfer genes from one population to another, as the hybrids cannot breed into either population. The situation often isn't initially symmetric. Sometimes hybrids can breed into one population and not the other, or either males or females can breed into one of the populations, but the other sex cannot.

Give it time, though, and hybrids cannot pass their genes on to either population. When hybrids cannot breed into the parental populations, we consider the two populations as different species. After more isolation, hybrids are no longer possible. Most of the genetics of how incompatibility forms are still mysterious, but we know of many genes that play a role. Many scientists research speciation, and we will someday soon understand the process in much more detail. For the purpose of this section, however, I will ignore the details of how genetic incompatibility forms and instead focus on how the necessary isolation that leads to incompatibility can arise.

There are, generally speaking, two models for how genetic isolation happens. The *allopatric* model where groups diverge when they are geographically isolated until they diverge enough to become separate species, and the *sympatric* model where groups split up for some other reasons while still in close contact. Some models mix the two, but we can consider just these two extremes without losing generality.

With modern modes of travel and the speed and quantity at which global travel happens, the geographical isolation of human groups will not be in our future here on Earth. Allopatric speciation doesn't seem likely in this scenario. Even if travel time between locations in the solar system is measured in years or decades, it is not enough to sufficiently limit gene flow to cause genetic isolation. We need much larger distances. If, however, we spread through the galaxy, *then* travel time will be measured in hundreds of thousands of years, and this appears to be sufficient from what we have studied of speciation in nature. It is not uncommon for species that diverged a million years or so ago to be able to hybridize, but sterile hybrids are then common. Give it a million years, and we might have many different species descendants from modern-day Homo sapiens.

We could end up with several species that are genetically incompatible, but it is also possible to end up with one large species where genes can flow freely between neighboring populations but where distant groups are genetically incompatible. It is a phenomenon we see on Earth called ring species. A ring species is a species that has expanded around some geographical barrier, like an ocean or a mountain. From some original location on the edge of the barrier, it spread in both directions. In both directions, some genetic divergence happens, but all populations can interbreed with their neighbor populations because they are genetically sufficiently similar. Except that when the expanding species meet at the other end of the barrier, the two branches have accumulated too many differences to interbreed. There are few differences between the neighbor populations along the barrier, but they add up, so at the extreme, too many changes have accumulated; where the extremes meet, there are enough changes for genetic isolation.

Genes can flow freely from one population at the extreme point of the barrier to the other population there, but only as long as they go through populations all the way around the ring; they can't go there directly. It is still a single species when we consider all the groups, but if we didn't have the full ring, we would classify the two neighbors at the extreme point as different species. Biology is always messy, and a clear species concept is unfortunately not something we have. With ring species, we have both one and two species in one. As humanity expands through the galaxy, colonists might diverge along the way. When they meet other groups of colonists, they will encounter new species, even though they can still exchange genes with the populations they recently diverged from. And populations they split from can exchange genes with the ones they in turn split from, and so on back to the most recent common population of the two groups of colonists, that we now consider as different species.

What are the chances of sympatric speciation, i.e., one that happens between humans not separated in space? It is hard to say, but the chances are probably quite small. It is not uncommon in species that can self-hybridize, like some plants. A single mutation—in plants, often a duplication of the entire genome—can make one individual incompatible with the entire species. This is a dead-end for an organism that needs sex to reproduce, but if the mutant can make clones of itself, it is instant speciation. This doesn't happen in animals. We cannot reproduce by cloning ourselves, so any mutation that makes us incompatible with the rest of the species means the extinction of that genetic line.

Still, sympatric speciation is possible in animals as well and has been observed. A scenario that can lead to it is this: imagine some traits with two or more adaptive genetic variants that are selected for. This can be two different preferences for food or sexual preferences for mate choice, or something similar. The genetics behind it can be a single gene, where having two identical alleles is better than having two different alleles. If you have both alleles for one or the other of the adaptive traits, you are fit, but if you have one of each, your trait falls somewhere between the two optimums—so you are less fit. If there is such a trait, it is beneficial to select mates with the same trait. For example, if there are two different feeding strategies— a somewhat contrived example could be eating grass or leaves. You need two identical alleles for either strategy, while if you have two different alleles, you are gnawing on tree bark and starving. I know it is a ridiculous example, you won't find such silly bark-eating and starving animals in nature, but I hope you see the problem for hybrids of two diverging strategies. If you have two identical alleles, you want to find a mate with the same two alleles. That way, all your offspring will have two identical alleles as well and thus be fitter compared to offspring of a mating between the two traits—the starving bark eaters. Over time, the gene flow between the two traits will diminish, and the original species will split into two. In animals, this is less common than allopatric speciation, but it does occur.

Could it happen in humans? Possibly, of course, but I find it unlikely. We don't have different feeding habits like in the example. We have different food preferences, but those are cultural rather than genetics. The most likely trait, I think, where we could see divergence is sexual selection, and even here, I find it highly unlikely that we will see sympatric speciation. While we have a wide variety of human appearances between ethnic groups, we don't see strong signals of exclusively breeding within groups—people mate between ethnic groups every day. We have cultural and religious groups that discourage mating with outsiders of their groups. However, we still see plenty of evidence of gene flow in and out of them, when we examine their genetic compositions. We also see gene flow when we examine the DNA of relatives of Homo sapiens, the Neanderthals and Denisovans. We are quite promiscuous, and we seem willing to mate with anyone we can get in contact with. This is not to say that sympatric speciation is impossible in our future; I just cannot imagine how it would come about.

When it comes to speciation because of separation, we have not only separation in space but also time. The genetic composition in humans will change over a million years, so humans today and humans a million years from now could be classified as different species. It is not guaranteed that gene flow between current humans and such future humans would be impossible. We see, in nature, that hybrids are possible between species that have diverged more than a million years ago, which is a separation of *two* million years (one million on each branch since the split). But fossils of early humans that are a million years old are undoubtedly classified as different species, and perhaps we would be considered a different species from those future humans as well. Without any split into separate species, we will become a distinct species in time; we just have to wait.

There are several ways that we can turn into a different species in a million years. Time alone will do the trick, and while there might still be only a single species of humans in a million years, it will be humans that are different from us. If we spread out through the stars, and faster than light travel turns out to be impossible, we will most likely diverge into multiple species.

Unnatural Selection

Is our future evolution through natural selection at all, though? So far, I have written about natural selection and our evolution in what I expect our future to look like as we adapt to our new environments. But the technology we currently have, and the more advanced technology we will have in the future, could change everything. I have argued that it is impossible to eliminate evolution as long as there is a genetic variation that we pass from generation to generation, and this is true. However, we are not forced to evolve according to nature's whim. We can guide our evolution through direct genetic manipulation.

Instead of having natural selection choose which genes should survive in the future, we can do so ourselves. This is not science fiction; we already do this today. We commonly screen embryos for genetic disorders and decide whether these genes should be added to the next generation's gene pool. The screening happens in pregnant women, and many have abortions based on the results. It happens with in vitro fertilization, when we do a genetic screen before an embryo is implanted. We do it today to screen for, for example, Huntington's disease and Down's syndrome.

Many will object that we should not screen for traits and select children based on the result, but nevertheless, it happens every day. When parents have the choice between a child with a genetic disease or to wait for one without, many naturally choose to wait and try again. It is easy to have a moral stand on the abstract scenario, but when it comes to your own children, you want what is best for them. You want them to be healthy and able to live a good life. You do not wish to bring into this world a child you know will live, perhaps, a short life in constant pain. I would go so far to say that I find it immoral to do so if you can avoid it. To me, avoiding unnecessary pain is a morally superior choice.

We are talking, obviously, about severe disorders. We do not screen for harmless things like myopia. But when genetic testing gets cheaper, and when we know more about the genetic basis for more diseases, we will inevitably screen for more traits. We will not jump directly from screening for Down's syndrome to testing for eye color, of course, but we will slowly, step by step, screen for, and select offspring based on, specific genes. We screen for severe disorders today. Tomorrow we will screen for less severe diseases, but still, disorders that will substantially reduce the child's quality of life. The day after tomorrow, we might screen for traits that, while not leading to debilitating diseases, are still severe enough to make the testing worthwhile. Would the next step then be to select for genes that lower the risk of diabetes, cancer, or hypertension? And once it is the norm in society to screen for disorders or slight increases in risks of a disease, how large a step will it be to select for positive traits? Higher intelligence or attractiveness, perhaps? We will get to the point where this is doable with the pace of improvements to genetic science and technology we see today.

I know it looks like a slippery slope argument, but that is not my intention. We are not in a situation where, if you say A, then you must also say B. We continuously choose what we will use our genetic knowledge for, and we can stop at any point if we feel uncomfortable. I just don't think that we will. I don't think we will stick with an ethic that says that it is better to have a child that suffers, or cannot match his or her peers in essential abilities, than it is to select embryos based on their genes. Parents, as a rule, want what is best for their children. If they can prevent harm or give them a boost in life, they will do so. I would be surprised if we are not there in less than a century. The time it takes before society will accept the practice, and the time until it becomes the norm to screen for desirable traits, will lag behind the science, maybe by decades or perhaps by centuries. I expect the cultural acceptance will follow, and probably quickly, once the technology is there.

If we select for traits based on screening, we only have the variation already in the parents to work with. The embryo will be a combination of these (plus a tiny number of mutations), and we cannot pick and mix genes from the entire human species. This is a limitation we can get rid of if we explicitly modify genes instead of selecting them based on screening.

Modifying genes might sound like science fiction as well, but it isn't. The first genetically modified children are already born. The first, a pair of twins, was born in November 2018. They were modified to increase their resistance to HIV. The doctor doing the gene-editing was sentenced to jail afterward, and rightly so—no one knows what the consequences of this editing are, and we need to know the ramifications of our actions before we experiment on humans. However, the technique used to do this, called CRISPR, is used in labs all over the world, and nothing except ethics committees and legislation prevent others from modifying embryos.

The primary obstacle to modifying embryos, to eliminate their risk for disorders and diseases (and probably someday to select for other traits), is not the gene editing. It is that we do not know what the consequences of our fiddling around in the genome are. We do not yet understand the myriad ways in which genes interact, and we cannot predict the implications it has on the entire organism if we change even a single gene. It is common to think of genes as doing only one thing. This gene gives you blue eyes, that gene reduces your risk of cancer, and that one over there increases your intelligence. But this is rarely the case. One gene does many things and in combination with many other genes. If we take a variant of a gene that increases your risk of cancer and replace it with another variant, we don't know what consequences it has for the other functions the gene has. It is not likely to be a problem, the working gene is already functioning fine in other people, but the point is that we do not know the outcomes of our action. It seems evident that it would be better to replace a cancer-causing gene with one that protects you from cancer, but we could be making something else worse. Reducing the risk of cancer by 0.001% at the cost of guaranteed worse eyesight, perhaps.

We cannot ethically modify the genomes of humans if we don't know the result. We have the technology to do it, but not yet the sufficient knowledge to do it ethically.

Each day, our understanding of gene functions grows, and we work out which traits our genes are involved in modifying, how they interact, and what the consequences are of one allele rather than another in a gene. In the last decade, we have seen an explosion in studies that elucidate gene functions and gene interactions. New technology allows us to sequence gene variation and to directly measure which genes are turned on and off in each cell type. These techniques are doing for genetics what the invention of the telescope did for astronomy—our knowledge is greatly enhanced, and we keep discovering more. How our genes function will not be hidden from us for much longer; we will soon have an atlas that tells us, in great detail, all the functions of a gene and the implications for replacing one variant with another. When we do get such an atlas, expect gene-edited babies to be commonplace in a few decades to a century after.

If we can put together each new generation precisely as we please, then you can rightly argue that natural selection is no longer active. The genetic composition of the next generation is not determined by who got to reproduce and who didn't, but instead, how parents choose to compose their offspring's genes. There will still be evolution and selection, though, just a different kind. What we consider optimal genes for our children will, in part, be a cultural preference. Some traits will be shared by all cultures, of course, such as good health for our children. But other attributes will follow specific cultural values and fashions. Especially if we cannot optimize on all parameters at once.

If, for example, you can either get musical ability or an aptitude for long-distance running, there might be a cultural/fashion aspect to what parents will choose.

Cultures and ideas evolve and compete in ways that can resemble natural selection. The idea that more people pick up and spread around has a selective advantage over more easily forgotten ideas. Ideas mutate and change into related ideas, and they can transform into something unrecognizable over many jumps from one mind to another, as anyone who has played Chinese whispers will know. We have mutations and selection and thus the evolution of ideas, cultures, and fashion. If we start modifying our genes according to our whims, then it is this cultural evolution that will drive us forward as a species.

Birth of A.I.

But the future might not belong to biological humans at all. Maybe it belongs to our "offsprings," thinking machines.

In many areas, we have seen a long slow growth in artificial intelligence performance, a performance that long stays much below the human level, until suddenly we have a machine that can compete with the best humans, and very shortly afterward, we have machines with superhuman abilities. We had computer programs that might beat normal players for both Chess and Go but would never stand up against champions. Until we got Deep Blue for Chess and AlphaGo for the game Go. These artificial intelligences can defeat the world's best players, and the games are not close calls. These are games with a small set of rules (but a vast number of combinations to consider when playing), and you could object that such games match the strengths of machines more than humans. But we cannot even hold our own in something like Jeopardy, where the A.I. Watson defeated human champions. This is a game where you must interpret hints and have a well of general knowledge. It seems we can build machines that are smarter than ourselves, at least when considering limited application areas, machines known as *narrow sense artificial intelligence*.

We also have some successes with building artificial intelligences that can learn by themselves. AlphaZero, created by the same people that made AlphaGo, can learn how to games by itself, and after a few hours' training can compete with, and often best, artificial intelligence built for that specific game. Let that sink in. We can build machines that can defeat the best humans, and we can build machines that can teach themselves how to beat the machines that can beat us. This is both awe-inspiring and a little frightening.

Still, AlphaZero only plays a limited type of games, so although it can learn more than one game on its own, it is still a narrow-sense intelligence. A machine that can handle a broad range of general tasks, in the real world, an *artificial general intelligence*, is not something we have yet. We do not know how to build machines that can learn and adapt at the level we humans can. Not yet! But people are working on it, and if we scoff at the chance of getting there, we are doing what people did with Chess and Go computers. We will claim that it can never happen, and then it does. If the speed at which we go from a decent chess player to a superhuman one repeats for general intelligence, we could quickly go from a simple artificial intelligence to human-level artificial intelligence, to superhuman intelligence. We don't know how to get there yet, but a decade ago we didn't know how to get a computer to play world-class go. We might not get there in centuries, but we know that intelligence is possible to create—we are here, after all, so it is naïve to expect that we will never get there. Eventually, we will.

There might be a limit to how complex an A.I. we can build. Human intelligence itself is limited, after all. But maybe we can create a machine that can create artificial intelligence, and it can create an intelligence superior to itself because it is better at it than we are. We can build a narrow-sense A.I. that is better than ourselves, so what should prevent us from making a machine that is better at building A.I.s than we are? Then that intelligence can create an even better intelligence. The first many iterations of this might result in intelligence inferior to our own, but computers can process and manipulate data much faster than us, so those generations could pass in a handful of years, maybe months, maybe days, or perhaps even minutes or seconds. Then the process reaches a point where the results are at the human level and imminently after that, intelligent machines better than we could create ourselves—and machines immensely more intelligent than us. We will no longer be the most intelligent species on Earth.

I don't fear a war between humans and machines, Terminator or The Matrix style. There are still dangers to creating artificial intelligence, though. If their goals and ours do not align, and they are vastly more intelligent than us, then it doesn't take much imagination to see who gets their will. If there is no conflict between their goals and ours, this is not a threat to us—they do their thing, and we do ours. If there is a clash between what they want and what we want to achieve, we will be prevented from doing what we want. If we become an obstacle to their goals, and they do not have as a goal to preserve us, then they could decide to eliminate humanity. Creating superhuman artificial intelligence is a potentially existential threat, and we need to get it right. In my eternal optimism, I believe that we will.

If artificial intelligences make copies of themselves, with modification, we will see selection and evolution of these new lifeforms. The time scale for their development will be vastly faster than for biological life, however. It is an evolution where each generation is explicitly built to adapt to the existing environment. It will not rely on random mutations. The time between generations can be far shorter than between human generations. Nothing prevents an A.I. from designing the next generation within minutes of going online itself. If machines replicate faster, adapt faster, and are already more resilient to environments where we cannot survive, it will be these machine children of humanity that will colonize the galaxy and not ourselves. We might live among them for many millions of years to come, but they will be the dominant species, not us.

You might think this is a bleak view on humanity's future, but think about it this way: they are our children, albeit in a different medium. If we manage to endow them with the goals, dreams, and qualities that we possess, why should we think of them as much different from our biological offspring? It won't be genetic offspring, but if you hope that your genes will continue down the ages, I am sorry to tell you that the chances are slim. Each generation, some of the genes you have contributed to the human gene pool are likely to be lost. The chance that a particular allele you carry will survive to spread through the species is the same as the chance that a new mutation will survive. Think of your specific allele as the mutation and consider its probability of getting fixed in the population. Earlier in the book, we calculated this probability to be one in 14 billion—although that chance depends on the number of people in each generation. If the human population size stays at seven billion, this is the result we get. If the population size grows, which I expect it to do over a million years if we expand into space, then the chance is much smaller.

Some of your genes *might* spread to all of humanity in the future if the allele copies you carry get fixed. The chances are tiny, though; you are competing against all of humanity in this lottery. All genes that do not replace the rest of humanity's copies will be lost. Even those distant descendants that can trace their pedigree back to you will likely not carry any of your genes; they also got their genes from their other ancestors, and the contribution from all of humanity swamps your contribution. You will have contributed the same number of genes to the future biological humanity as you have contributed to the artificial intelligence humanity.

If we build machines much like ourselves, and these machines colonize the universe, we will have descendants of both flesh and steel. The machines will be far more numerous and widespread, but this does not have to mean that the biological humanity does not prosper and join in the spread among the stars. Both groups will be our descendants, and, hey, maybe they will even get along.

Ghosts in the machine

If you don't mind a mechanical destiny for humanity, but you would like it to be actual humans, there is also a possible future for that—merging with the machines.

Cyborgs walk among us. Not the science-fiction type, but people with artificial hips, people with pacemakers, people with prosthetics. It doesn't appear that we are afraid of replacing our human parts with machine parts. At least not on a minor scale. And it is not just body parts we are willing to replace or enhance. We add machine parts to our brains. Brain implants are used to treat Parkinson's disease, to allow the blind to see, and the deaf to hear. Treatments that involve inserting chips in your brain are still rare and limited, but as we learn more about the brain, and as our information technology improves, we will see it more of it, and we will see it for a wide variety of disorders.

As with genetic engineering, we are limited by our understanding of the underlying biology, but knowledge of neuroscience moves at least as fast as genetics. So expect the possibilities that are open to brain-implements to grow quickly in the immediate future. At first, we only use it to treat disorders—the same scenario as with genetic engineering—but soon, we will use it to enhance ourselves. If you have a poor memory, as I do, you might want to get a chip that helps you store and retrieve information. If you are poor at math, why not implant a calculator? It sounds outlandish now, but how different is it really from having a calculator on your phone? That app enhances your math capabilities, it just has a worse user interface than merely thinking about the calculations you want to do. If implants are cheap and easy to get—with some delivery mechanism that doesn't involve brain surgery—it is easier to have your apps directly in your brain than on your phone. How about instant access to all the information on the internet? You could get the information from Wikipedia as quickly as you recall memories.

We can use our technology to enhance ourselves and transcend our biological limitations. We can adapt to our environment much faster with technology than we can with genetics. We can replace body parts with artificial parts. We can get artificial organs instead of waiting for donors, and we might be able to achieve virtual immortality by replacing parts whenever necessary. Where it truly gets interesting is when we can replace more and more of our brains. If we replace one small piece at a time until there is nothing of the original brain left, we end up with a situation like Theseus's ship: If you replace parts of a ship, piece by piece, year by year until there is nothing of the original material left, but there is still a ship that looks exactly like the original, is it still the same ship? This happens with your body as you are reading this. Cells die and are replaced all the time. The atoms in your body come and go. The configuration of your body is what remains, not its parts. If we replace your brain piece by piece, while your essential personality remains intact, is the resulting person still not you? If not, why isn't the same the case when we replace a cell at a time?

If we get to the point where we can modify ourselves to this degree, we can build bodies to match our environment and needs, and we can enhance our intelligence as we improve our brains. We will be able to do what the artificial intelligence in the previous section could. Enter a loop of improvements where each iteration builds a new and better version of ourselves. There would not be much difference between such a transcendent human and an artificial intelligence build from scratch, except perhaps in psychology and personality. The future trajectory for both would be the same. If we both develop artificial intelligence and enhance ourselves to this point, there probably wouldn't be much difference between them and us.

In this scenario, biological reproduction will, at most, be something that happens at the very early stage of our lifetime. We cannot reproduce if we have replaced our sexual and reproductive organs with artificial components that do not serve exactly the same functions. And why wouldn't we? We only need them to reproduce, and we won't necessarily want to reproduce indefinitely in what is in practice an immortal life. Our sex drive is strong in our biological form, and we would all hesitate to get rid of our sexual organs if asked today, but that drive is something we can remove when we update our brains. We might end up with a species that reproduces in its early life, say within the first hundred years, and then changes into a form that instead pursues arts and science and explores the universe. It is not much different from today, where we typically reproduce before we reach fifty, and grandparents have other things to do. It is the same as one of the scenarios we explored in the Demographics chapter, except that this time we have achieved immortality by becoming machines.

Biological reproduction could also disappear entirely, and our descendants become wholly artificial. We could create copies of our brain patterns and transfer them to new synthetic bodies, or we could mix personalities from ourselves and one or more partners to produce offspring. Instead of a sharp change between humans and a machine intelligence phase of humanity's future, we would have an intermediate phase with part biological and part mechanical humans. The result would be the same. Our future would be a species of machines.

Evolution would not go away. There would still be variation, and some variants would be better adapted to the environment we would find ourselves in. But we can mutate at will and change at short notice. We can let ourselves be inspired at other's solutions to problems and imitate them—rewrite our own software based on both our personal experience and the experience of others. In essence, copy the 'genes' that we like from others whenever we need them. Bacteria already do this—it is called *horizontal gene transfer*—but we will now be able to do it as well. We will just be able to do it faster and better, with intelligence guiding the gene exchange. It is nice to think that we might get better at the evolutionary game than bacteria in the distant future. We would keep evolving, not with natural selection but with artificial selection, as a species very different from today's humanity, onward into the far future.

CHAPTER 7

Conclusions

All the predictions in this book are pure guesswork. I am not psychic. I have mainly listed tendencies that I expect to continue into the far future, and most of the world that I envision will be reached in a millennium or two, if not in a century or two. After that, I have assumed that the next million years to be more or less the same. Such a static future will, of course, not be the case. Our technology and culture will change rapidly, and so will the future fitness landscape.

The trends I have listed are those I think are most likely to be true far into the future. I am mostly optimistic and expect our future to be wealthy, urban, and mobile. I hope that we will collaborate and mix more. The trend is already there, despite the occasional jingoism. In the competition between nations, wealth is what we aim for. War makes us poorer, while trade makes us richer. We are prosperous when labor and goods are mobile and when there are few restrictions on trade—this is why we bother to make free trade agreements. Trade wars do not make anyone richer in the long run; actual wars can speed up innovation, but they leave the world poorer.

I expect that we will generally use our technology for good, and I anticipate that global wealth will continue to rise. While some will be richer than others, poverty will be a thing of the past. There will be haves and have-nots, and wealth will be unevenly distributed; I do not foresee a socialist utopia. Human nature, it seems, prevents this. There will always be those that seek wealth and power. Those are adaptive traits, and as long as they are, they remain in our society. In all probability, we will always be an unequal society, to a greater or lesser extend. Still, the technology that is at first only available to the rich and powerful will, in a generation or two, be available to everyone. History shows us this. We will all ride on a wave of improving science and technology, and where this wave takes us will determine our future.

In most of the book, I have talked about natural selection and biological evolution, but I do not believe that those forces will be the main drivers of our future as a species. They are slow in changing us, even though they should be more efficient now than in our past; they are guided by nature and will move us in directions we might not want to go; we can do better than this. I expect that frequent gene editing lies in our future and that cultural evolution, affecting how we choose to edit genes, will replace biological evolution. For good or for worse, I do not think we will be slaves of random mutations and nature's whims for much longer. We will decide ourselves how our genetic future will be. And perhaps leave biology behind altogether. In my view, the last chapter describes the most likely future we can expect.

All my guesses can be invalidated as early as in the next decade or the decade after, or any point in the millennia that follows. A nuclear war will seriously mess up the premises for my speculations. Doomsday devices like bioweapons or gray goo, nanomachines that convert all matter into copies of themselves, could wipe us out. I have to assume something about the future to extrapolate our possible evolution from, so I pick one based on optimism. If you had hoped this would be a book about a dystopian future, I am sorry to disappoint you. Despite the doom and gloom in the news, the life of humanity improves steadily, each generation is generally better off than the previous, and I firmly believe that this will continue. I am optimistic about where evolution will take us, and I wish I could come along on the ride. I will not be here in a million years, but my genes might make it. Yours might as well. Perhaps they will meet someday.

Sources

Unless I have explicitly given a source, all numbers I have mentioned in the book comes from (https://ourworldindata.org) an excellent source for world wide and historical statistics.

Acknowledgements

I am grateful to Maria Izabel Cavassim Alves for many useful comments to earlier versions of this manuscript.

www.ingramcontent.com/pod-product-compliance
Lightning Source LLC
Chambersburg PA
CBHW052352220526
45465CB00003BA/1065